SETTING UP A TROPICAL
AQUARIUM

WEEK-BY-WEEK

SETTING UP A TROPICAL
AQUARIUM

WEEK-BY-WEEK

STUART THRAVES

FIREFLY BOOKS

A FIREFLY BOOK

Published by Firefly Books Ltd. 2004

First printing

Publisher Cataloging-in-Publication Data (U.S.)
Thraves, Stuart, 1970-
     Setting up a tropical aquarium week-by-week /
Stuart Thraves. _ 1st ed.
[208] p. : col. ill. , photos. ; cm.
Includes index.
Summary: Step-by-step and week-by-week guide to setting up and maintaining a tropical freshwater aquarium. Includes the answers to commonly asked questions and practical advice for every stage of an aquarium's development.
ISBN 1-55297-933-4
1. Aquariums. 2. Marine aquarium fishes. 3. Tropical fish.
I. Title.
639.34/2 22   SF457.3.T57 2004

National Library of Canada Cataloguing in Publication
Thraves, Stuart, 1970-
     Setting up a tropical aquarium week-by-week /
Stuart Thraves.
Includes index.
ISBN 1-55297-933-4
     1. Aquariums. 2. Tropical fish. I. Title.
SF457.3.T48 2004   639.34   C2004-903279-8

Published in the United States in 2004 by
Firefly Books (U.S.) Inc.
P.O. Box 1338, Ellicott Station
Buffalo, New York 14205

Published in Canada in 2004 by
Firefly Books Ltd.
66 Leek Crescent
Richmond Hill, Ontario L4B 1H1

Design, graphics and prepress: Stuart Watkinson
Photography: Geoff Rogers
Production management: Consortium
Print production: Sino Publishing House Ltd.

Printed and bound in China

The Author

Stuart Thraves has had a lifelong fascination with fish. This led him to study them at Sparsholt College in Hampshire, England. Stuart's interest in fish has also developed into a favourite hobby – diving – which he enjoys pursuing around the UK and in the Red Sea. Today, he is Brand Manager of a leading aquarist supply company. His day-to-day experience of the hobby keeps him abreast of the latest developments and provides an ideal platform for writing this book.

INTRODUCTION

We are constantly fascinated with other worlds. Apart from outer space, the only other domain we have yet to completely explore is the underwater world. It is a world of sumptuous variety, from the swirling sun-drenched waters of a coral reef to the tranquil shadows of an Amazon stream trickling slowly beneath a canopy of heavy-leaved tropical trees. These are just two examples of many thriving habitats in which plants, fish and invertebrates live together in aquatic harmony.

Since the great age of discovery and invention in Victorian times, we have tried to bring this fascinating spectacle into our homes. The first aquariums were simple boxes with glass fronts in which temperate fish species were kept. With little appreciation of their needs, the fish soon expired and had to be replaced regularly. This approach would be totally unacceptable today with our regard for animal welfare. A greater understanding of how natural cycles work and the technology to sustain them now offers fishkeepers the exciting opportunity to create and maintain aquariums that will thrive for many years.

This book follows the sequence of setting up and running a tropical freshwater aquarium for the first twelve weeks of its life. At logical points along the timeline of practical progress there are profile sections that feature extended selections of suitable plants and fish. Throughout, the emphasis is on explaining how vital life-support systems work, including the basic science behind the processes they depend on. Unlike many hobbies, where learning can be phased and levels of difficulty matched with emerging ability, fishkeeping requires the immediate grasp of a crucial technique — how to turn a transparent box of water into an environment that will sustain life from day one and beyond. Fishkeepers are pet owners, with all the responsibilities that title entails. How you control that environment dictates whether your display will thrive or deteriorate into an unsightly mess. This book will give you a head start in the quest to be successful.

Above: *Two clown loaches probe the substrate for food as a lone zebra danio hangs back from the other members of its shoal and a colourful male cockatoo cichlid seeks his female companion.*

CONTENTS

DAY 1
Building the system

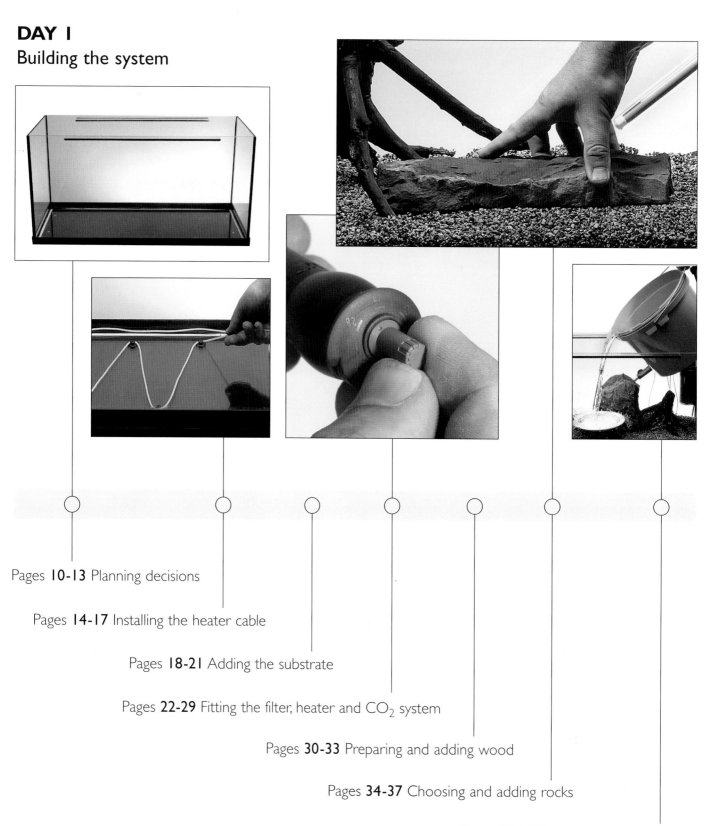

DAY 2
Planting up the aquarium

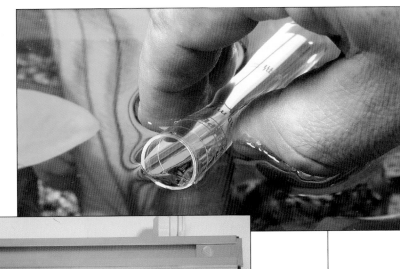

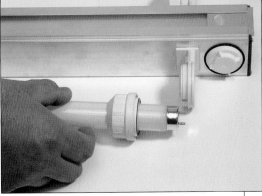

CONTENTS

WEEK 3
Bringing the aquarium to life

WEEK 5
Increasing the fish population

WEEK 8
Towards a stable system

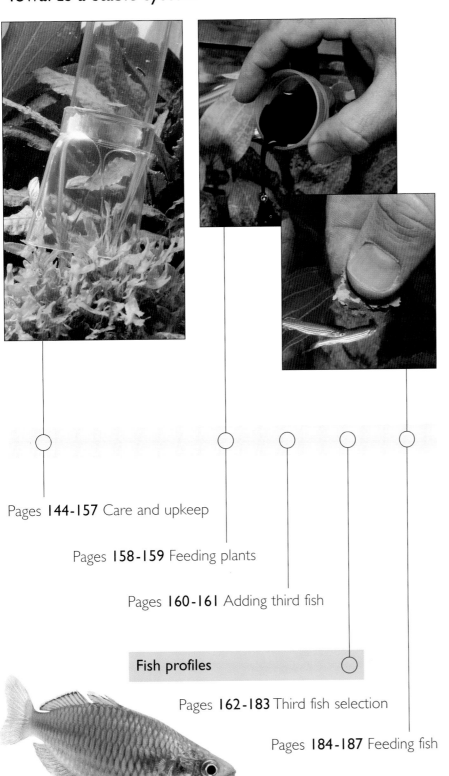

WEEK 12
The finished display

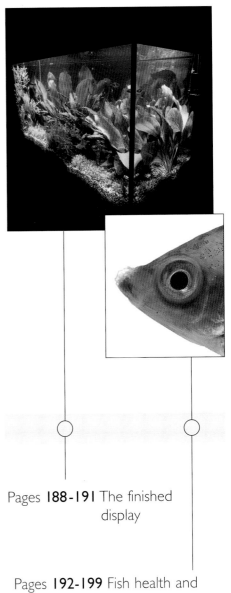

FIRST THOUGHTS AND DECISIONS

When you have decided to set up a tropical freshwater aquarium, your crucial first decision is where you would place it in your home. Clearly, it will dictate the size and shape of tank you can consider.

Choosing a location

There is usually a place in every home suitable for an aquarium. Its location can have a great influence on how successful an aquarium is. It might mean changing the layout of an entire room to accommodate the aquarium in its best position. A good idea is to draw the layout of the room on paper and sketch in the likely positions of the aquarium. This could save hours of moving furniture and then finding out that everything simply will not fit!

Choose a location that will present the aquarium as a focal point in a room where people spend the most time relaxing in the evenings or at mealtimes. A favourite place for an aquarium is as a room divider, enabling the aquarium to be viewed from both sides. It can also be set into a wall to provide the classic 'living picture'.

Once you have chosen a suitable location, check that the tank support is strong enough. This includes checking that any pre-assembled cabinets have all the bracing pieces fitted. The aquarium featured in this book measures 90cm (36in) long, 38cm (15in) from front to back and 45cm (18in) high. Such an aquarium full of water will weigh about 150kg (330lb)! Also check that the floor can support the weight. The essential rule is that the aquarium should run across several floor joists and not along a single joist. If in doubt, consult a qualified builder for advice.

Finally, check that you can get the tank to the room in the first place. Aquariums are not flexible or available flat-packed and are generally very heavy, awkward and fragile to handle.

Choosing a tank

When you have a suitable location in mind, choosing a tank is your next decision. There are many shapes and sizes available today and it is a good idea to seek expert guidance from your local retailer.

Think of the fish

The first golden rule to consider when buying an aquarium is that it must provide the correct conditions for your fish to survive. Like us, fish 'breathe' oxygen and 'exhale' carbon dioxide. These gases are dissolved in the water and get in and out at the surface. Therefore, any aquarium must provide a large surface area to allow sufficient exchange of gases.

This key point greatly influences the optimum design for an aquarium. Fortunately, the best design from a 'life-support' point of view also looks pretty good in the living room! Since we are aiming to create a living picture that

Siting the aquarium

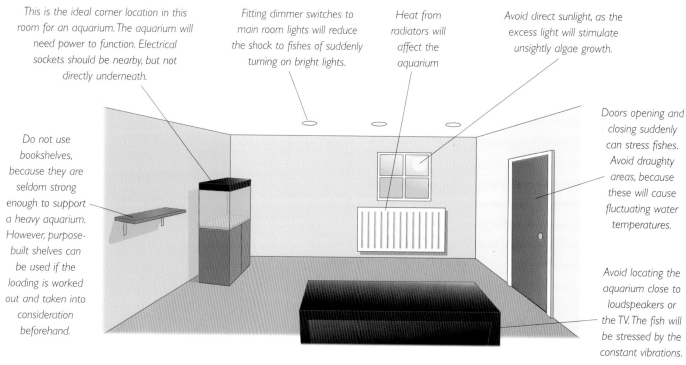

This is the ideal corner location in this room for an aquarium. The aquarium will need power to function. Electrical sockets should be nearby, but not directly underneath.

Fitting dimmer switches to main room lights will reduce the shock to fishes of suddenly turning on bright lights.

Heat from radiators will affect the aquarium

Avoid direct sunlight, as the excess light will stimulate unsightly algae growth.

Do not use bookshelves, because they are seldom strong enough to support a heavy aquarium. However, purpose-built shelves can be used if the loading is worked out and taken into consideration beforehand.

Doors opening and closing suddenly can stress fishes. Avoid draughty areas, because these will cause fluctuating water temperatures.

Avoid locating the aquarium close to loudspeakers or the TV. The fish will be stressed by the constant vibrations.

Right: All tanks should have a large surface area to allow oxygen to be absorbed and carbon dioxide to leave the water. Cube-shaped aquariums provide interesting viewing aspects, along with an acceptable surface area in relation to depth.

Below: A rectangular aquarium is a traditional choice and the shape we have chosen to set up. This tank is 90x38x45cm deep (36x15x18in) and holds 150 litres (33 gallons) of water.

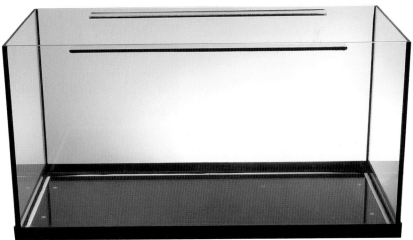

Above: Fishkeepers today have a wide choice of aquarium shapes and sizes. This curved tank on a purpose-built stand is supplied with lighting and filtration equipment included.

Left: A well-furnished aquarium can become a fascinating focal point in any room. But bear in mind that to keep it looking its best means carrying out regular maintenance, including water changes. Is there a water source nearby or will you have to carry heavy buckets back and forth each time?

recreates an underwater scene, the most effective and compelling format is a rectangular aquarium that provides a 'widescreen' landscape view. This is ideal for the fish because it offers a large surface area in relation to the volume of water in the tank.

This principle effectively rules out the tall column-style aquarium, which has a relatively small surface area. Only a few fish could be kept in this style of aquarium and would need a great deal more care than they would in a landscape format aquarium.

'Permissible' variations on a landscape theme include the bow-fronted aquarium — in a rectangular or corner shape — as well as six-sided aquariums. Also, compact cube aquariums are popular and ideal for small rooms.

Buy as big as you can

Aim to buy the biggest aquarium you can afford and accommodate. The simple reason for this is that when keeping fish the larger the volume of water you have to play with the easier it will be to control the waste products the fish produce and hence provide a stable environment. Although there are effective filtration systems available, the larger the volume of water you provide the better it will be for your fish in the long term.

Final checks before buying

Once you have decided on the aquarium that suits your requirements, ask yourself the following questions as a mini checklist to avoid possible problems in the future. How much access to the aquarium is there once the lid is on? How easy will it be to carry out regular maintenance? If the aquarium has a built-in life-support system, is this simple to maintain? Always check the glass of the aquarium and reject any with slight chips, since these are potential weak points. Even a compact aquarium holds a great deal of water that, together with the plants, rocks, equipment and fish, will make a fine mess of your living room floor if it shatters!

Above: Odessa barbs are a shoaling species, so keep them in groups. Give them swimming space, as well as areas of dense vegetation that provide hiding places and look decorative and natural.

Right: Cube aquariums provide great opportunities for creative planting. In this tank, a black background has been applied to the rear panel so that the display can be viewed from three sides. The heater and filter occupy one corner.

The light-dark cycle in the aquarium

■ Oxygen ■ Carbon dioxide

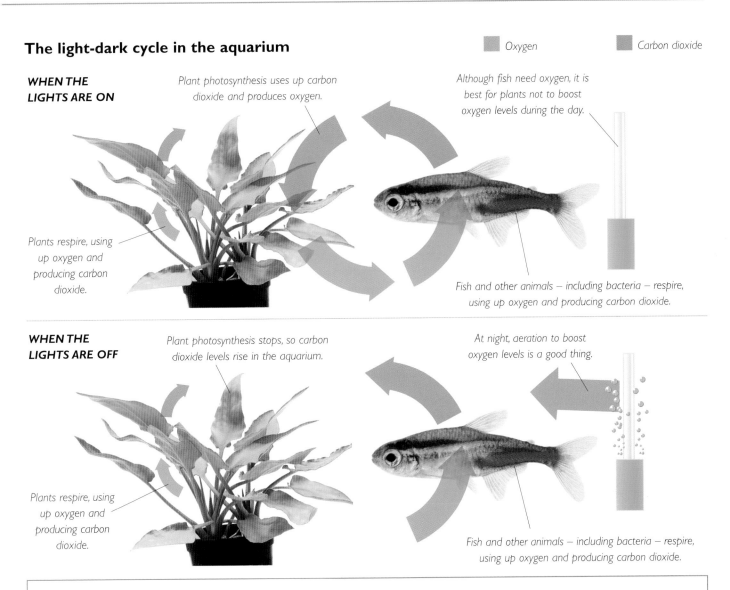

WHEN THE LIGHTS ARE ON

Plant photosynthesis uses up carbon dioxide and produces oxygen.

Although fish need oxygen, it is best for plants not to boost oxygen levels during the day.

Plants respire, using up oxygen and producing carbon dioxide.

Fish and other animals – including bacteria – respire, using up oxygen and producing carbon dioxide.

WHEN THE LIGHTS ARE OFF

Plant photosynthesis stops, so carbon dioxide levels rise in the aquarium.

At night, aeration to boost oxygen levels is a good thing.

Plants respire, using up oxygen and producing carbon dioxide.

Fish and other animals – including bacteria – respire, using up oxygen and producing carbon dioxide.

Attaching the background

If you intend to attach a flexible plastic background to the outside glass, remember to do it before you locate the aquarium in its final position. (In our step-by-step sequence, we have not fitted a background until the later stages so that we can more clearly show the practical sequence of setting up the aquarium.) There are many background options available, ranging from black and blue to printed designs that simulate planted scenes, rocks, tree roots and even historical ruins. Cut the background to size before fixing it on. If you are using a picture background, trim it to show the best part of the design.

① Stick adhesive foam pads at the top and bottom corners of the aquarium. Depending on the size of the tank, you may need additional pads in between.

② Apply the background firmly to the adhesive pads. This backing is double-sided, with black showing on the aquarium side. The reverse blue side provides another option.

BUILDING THE SYSTEM

The first day of the setting up process will be dominated by positioning the tank, installing the equipment and introducing the first of the hard landscaping decor into the aquarium. However, by the end of day one, you will begin to have a good idea of how the display will look.

Before you start assembling the equipment you have bought from the aquarium shop, make sure you have the following tools to hand:

A long spirit level
Screwdriver
Sharp scissors
Nail or scrubbing brush
Water jug
A couple of old towels
A four-way electrical extension
A plug-in 24-hour timer
Double-sided sticky tape or sticky pads for attaching the background (see page 13)
Cable ties
A 10-litre (2-gallon) plastic bucket

The bucket is one item that you will need for the entire life of the display, for transporting both clean and waste water. It is also the ideal place to store all your aquarium maintenance equipment. It is a good idea to choose one with clear markings so that measuring out the water capacity for your display will be easy. Alternatively, you can put your own markings on the outside.

You may also like to set aside some dry clothes in case your sleeves get wet or you are splashed during the setting up process.

Levelling the tank is the first vital step; even a slight discrepancy will be obvious once you add water to the aquarium. Use a long spirit level. If necessary, rest it on a straight wooden batten.

INSTALLING THE HEATING CABLE

Once you have chosen your aquarium and decided where to put it, you can begin the enjoyable task of setting it up. It is vital to proceed carefully in a patient frame of mind, following the step-by-step instructions provided here. Rushing to finish the display and failing to allow the emerging aquarium to mature at certain critical stages during the first few weeks will cause problems later on and dispel the good feeling you had at the outset.

Level and clean

When you have unpacked all the equipment and positioned the tank on the cabinet or stand, the first task is to make sure that the tank is level, both from front to back and side to side. This is crucial, not only because a 'sloping' water level will look unnatural and disturbing; but also because a tank that is not level is potentially unstable. Always adjust the cabinet or stand and not the tank. Placing wedges under a glass tank full of water will create stresses that may cause the glass to fracture. For the same reason, rest the tank on a plastic foam or polystyrene mat to compensate for slight irregularities in the surface of the support. (This is not

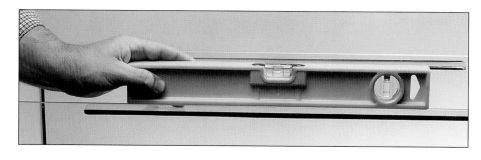

necessary for stands that support the tank around the perimeter of the base.)

The next job is to wipe the inside and outside surfaces with a damp cloth. The tank may have been in the shop for a few weeks and will have a coating of dust that should be removed. Use a clean cloth – ideally a new one – that has no chemical cleaning agents on it.

Installing the heating cable

With the bare tank level and clean it is now ready for the components that will create and maintain our living underwater scene. When installing equipment, it is vital that you work from the bottom upwards. The first item to go in is the heating cable that will provide a gentle warmth to encourage and sustain healthy root and

Above: Make sure that both the stand and tank are level by placing a spirit level along all the edges. Make any necessary adjustments.

plant growth. The cable will be covered by a layer of substrate and is impossible to install at a later date. Lay the cable across the inside base of the tank with a 3-4 cm (1.2-1.6 in) gap between the turns. Most models are supplied with suckers to anchor the cable; if not, use handfuls of gravel to hold the cable in place.

These cables are sometimes hard to find. The wattage should be no more than 25 watts to prevent root damage. Heating cables designed for other uses, such as terrariums, are too powerful and may burn and eventually kill plant roots in the substrate.

Below: Wipe the inside of the glass with a clean cloth to remove dust, otherwise it will create a film on the water surface. Also clean the exterior and remove any marks left by the packaging.

Right: Connect the low-wattage cable to the transformer and position this in a dry place outside the tank.

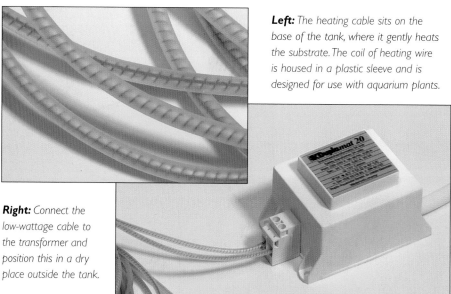

Left: The heating cable sits on the base of the tank, where it gently heats the substrate. The coil of heating wire is housed in a plastic sleeve and is designed for use with aquarium plants.

Nutrient flow in nature

The low-wattage cable generates a small amount of extra heat along the lines of the cable in the substrate. This causes thermal currents in the substrate that move the nutrients around and prevent the substrate becoming stagnant. This recreates what happens in nature, where the sun heats the riverbed or lake margin, making the natural substrate warmer than the water above it. Convection currents create a circulation of water that carries essential nutrients down into the substrate.

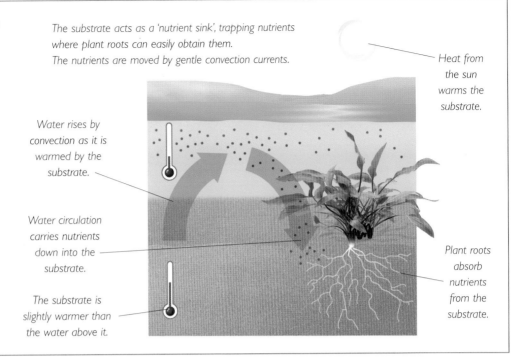

*The substrate acts as a 'nutrient sink', trapping nutrients where plant roots can easily obtain them.
The nutrients are moved by gentle convection currents.*

Heat from the sun warms the substrate.

Water rises by convection as it is warmed by the substrate.

Water circulation carries nutrients down into the substrate.

The substrate is slightly warmer than the water above it.

Plant roots absorb nutrients from the substrate.

Below: *Install the heating cable evenly across the aquarium, guiding it round the rubber suckers provided. Dampen these to help them stick to the glass.*

Keep the cable under slight tension as you feed it around the suckers.

ADDING SUBSTRATE

The next layer to install is the substrate. Many different substrates are available for aquarium use, but they all have one thing in common – they are all supplied unwashed. Placed directly in the aquarium, unwashed substrate would transform an aquarium into a cloudy maelstrom that any filtration system would have difficulty clearing. It is essential, therefore, to wash the substrate before use. Even apparently clean substrate will produce dirty water when first rinsed. To clean the substrate, simply wash it in several changes of tapwater in a bucket until the water is clear. To speed up the removal of heavy dirt, use boiling water, but be sure to stir this with a large spoon or similar implement to avoid scalding your hands.

The substrate used here is a porous ceramic material known as 'Alfagrog' and by other trade names. It is widely used in both aquarium and pond fishkeeping and is made by firing clay material at very high temperatures to produce inert rocklike fragments with a large surface area.

Medium-grade pieces are often used as a biological filter medium and even larger pieces as decor. Its inert nature makes it ideal as an aquarium substrate, and a small grade, with particles up to 3mm (0.1in) across, has been chosen for this aquarium. The structure is very open, allowing good heat circulation from the heating cable and excellent support for healthy root growth. The large surface area of the particles also boosts biological filtration in the aquarium because it fosters the growth of aerobic bacteria that 'process' organic wastes.

Use a clean plastic jug to pour the substrate carefully onto the base of the aquarium. Build up the depth to about 7.5cm (3in) and keep some washed substrate by for later use.

The role of the substrate

The substrate creates a natural 'bed' within the aquarium and supports any rocks and wood that you may add as part of the decor. Its main function, however, is to anchor and nourish the plant roots that

will grow through it. Unfortunately, most aquarium substrates, including gravel, do not contain any nutrients and cannot sustain plant growth on their own. The solution is to add a food source to the substrate to ensure healthy plant growth.

The most common of these longterm plant food supplements is laterite, a clay-based reddish additive that provides

Alfagrog is an ideal inert substrate that has no effect on water chemistry. By mixing laterite plant food material into the lower layers, it will sustain the plants.

Below: *Using a jug, pour the washed substrate into the aquarium to a depth of about 7.5cm (3in), taking care not to disturb the heating cable.*

Above: *Wash the substrate thoroughly in a bucket. Do not try to clean it all at once; instead, place a small amount in the water and agitate it until the water runs clear.*

Adding nutrients to the aquarium

1 *Add laterite evenly over the entire surface of the substrate, scattering it from the container in a side-to-side pattern.*

3 *Crumble the supplied bacterial culture between your fingers and add a thin layer on top.*

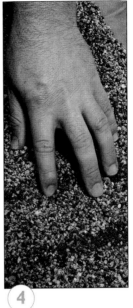

4 *Gently mix the ingredients into the substrate with your fingers and smooth out the top.*

5 *Add a final layer of substrate to create a finished depth of about 10cm (4in).*

2 *Work methodically from one side of the tank to the other for an even spread.*

essential iron to nourish aquarium plants. It needs to be used in conjunction with a bacterial culture that will release the nutrients in a form that plants can use.

The laterite will supply nutrients for several months in the aquarium. After this period, you will need to add fertiliser tablets close to the plant roots to continue feeding them (see page 159).

The featured aquarium uses a single layer of substrate with a seam of laterite to nourish the plants, but there are other substrate options worth considering.

Pea gravel

The most commonly available medium for the base of an aquarium is pea gravel, so called because of its rounded, pealike grains. This is available in a range of particle sizes, the most useful of which is 5mm (0.2in). This grade of pea gravel will have a fairly open structure and should not pack down too tightly over time, allowing oxygenated water to slowly circulate to all levels and prevent any area within the

Above: *A pea gravel substrate makes a suitable planting medium. Experiment with combining different grades for a more interesting effect.*

substrate bed from becoming stagnant and anaerobic (starved of oxygen).

If you have soft water as your main supply, the mineral content of the gravel will help to buffer the water (resist

change) and provide a stable neutral to slightly alkaline pH (see page 93). If your water is relatively hard and slightly alkaline to start with and you want to create more acid conditions, perhaps to suit certain fish, you may find it difficult to lower the pH if you use pea gravel, since the mineral content in the substrate will bring it back to neutral or alkaline pH values. In this case, consider using one of the completely inert substrates, such as lime-free quartz gravel, that will not affect your water chemistry in any way.

Quartz gravel

Quartz gravel is an inert granular material that is ideal for most aquarium setups. Colours can vary, with a dark sandy shade being the most commonly available. The grade best suited to aquarium use has a particle size of 1-3 mm (0.04- 0.12 in). This forms an excellent substrate that will allow good root penetration and growth to support a healthy plant. As with other gravel substrates, quartz gravel will need to be combined with laterite or similar nutrient material in the mid to lower levels to support plant growth.

Sand

When you visit aquarium shops you may see sand on the bases of some display aquariums. This will be silver sand, which is pale in colour and has very fine inert grains; do not use builder's sand, which has soft orange particles that will cloud the water. Silver sand is ideal for covering a heating cable on the aquarium base. Its fine texture allows good heat circulation from the cable to create thermal currents in the substrate. Do not use silver sand for anything other than a base layer because it quickly compacts down and will soon eliminate oxygen from the substrate, encouraging unwanted anaerobic bacteria to flourish. Always overlay the sand with gravel or other suitable substrate with a more open structure.

Lime-free gravel makes an excellent inert main substrate and is a good planting medium.

Above: *Here, the substrate is a mixture of gravels and small pebbles to emulate a streambed. It is surprisingly easy to alter the texture and colour of the substrate.*

Right: *If you choose a sandy substrate to emulate a particular natural habitat, disturb it regularly with your fingers to prevent stagnation. Clear away algae and debris with a clear siphon tube.*

Below: *Use silver sand as a base to distribute heat from a heating cable, but not as the main rooting substrate, as it soon compacts.*

Pea gravel is commonly available and cheap to buy. Mixed grades add variety and combine well with larger rocks and pebbles.

Red chippings add colour and texture. Chippings are available in various colours. For the best effect, choose ones that match the rocks in the tank.

Black quartz makes a good contrast to the more usual golden-brown colour.

Using soil in the aquarium

Although soil may appear to be an obvious choice as a plant-growing medium in the aquarium, beginners are not recommended to try it. However, if you develop your skills and decide to use soil as the main rooting medium, be sure to sandwich it between a layer of silver sand below and fine gravel such as quartz on top. Do not use garden soil, as it will contain unwanted nutrients, which could cause massive algal growth that would engulf your plants, as well as potentially harmful pesticides and other debris. The only recommended option is to use pre-bagged sterilised 'aqua soil' sold for use in plastic baskets for pond plants.

Soil has advantages in the aquarium because it has a high iron content and thus removes the need for additional iron fertilisation. Also, the aerobic bacteria that break down organic matter in the soil release carbon dioxide during respiration that aquarium plants can use as a food source. The main practical drawback is that if the soil layer is disturbed it will turn your display into 'mud soup' and you will need to strip down your messy aquarium completely and start again.

Soil can encourage unwanted algae growth, which can easily ruin a display. Beginners should use one of the many gravel options.

FITTING THE FILTER, HEATER AND CO₂ SYSTEM

The choice of aquarium filtration systems has never been so wide. All filters are designed to perform the same basic tasks – to remove solid waste from the water and provide a large surface area bathed with oxygenated water that will stimulate the growth of millions of beneficial 'cleaning' bacteria. A third task is to remove specific wastes or toxic products from the water with chemical filter media.

The internal power filter

Our system uses a compact internal power filter that sits discreetly in the back corner of the aquarium. A submersible water pump at the top of the unit draws water through the filter medium and circulates it back into the tank. The filter medium is a simple block of open cell plastic foam that admirably performs the

twin roles of trapping solid debris from the water flow and providing a large surface area for beneficial bacteria.

The filter body clips into a plastic cradle fixed to the glass with suckers. Starting with a new clean tank makes fixing suckers easy and reliable: suckers submerged for many months do become difficult to fix onto glass, especially dirty glass. Turn the filter in the cradle so that the return (cleaned) water flow is directed diagonally across the tank. This will set up an even circulation of water around the aquarium. The intake port of the filter (drawing in dirty water) is right on the bottom of the unit, so take care not to bury it in the substrate, as this will drastically reduce the filter's performance.

Regular maintenance is vital to ensure that the solid waste trapped in the filter

foam does not build up to a level that impedes the growth of filtration bacteria. A lack of oxygenated water will cause the bacteria to die off and the water quality will deteriorate, potentially causing lethal conditions for the fish. (See page 152 for practical guidance on filter maintenance.) If these filter units are well maintained, however, they should give many years of reliable service.

External power filters

As its name implies, an external filter is located outside the aquarium, with only minimal pipework visible inside the tank. Being generally larger than their internal counterparts, they have the capacity to house different types of media for mechanical, biological and chemical filtration. These are often housed in

An internal power filter

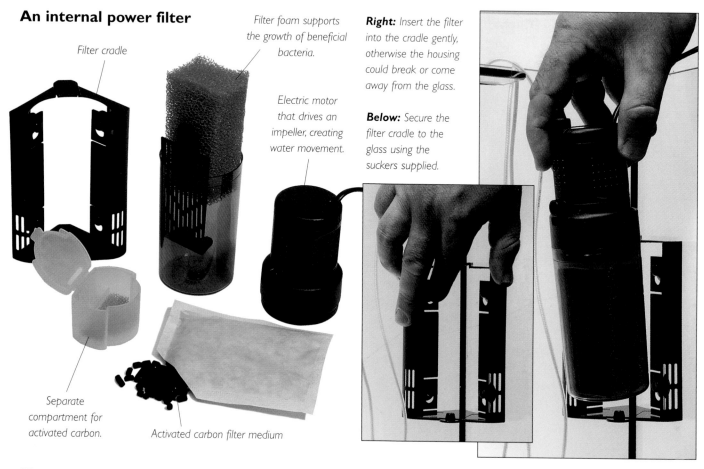

Filter cradle

Filter foam supports the growth of beneficial bacteria.

Electric motor that drives an impeller, creating water movement.

Right: *Insert the filter into the cradle gently, otherwise the housing could break or come away from the glass.*

Below: *Secure the filter cradle to the glass using the suckers supplied.*

Separate compartment for activated carbon.

Activated carbon filter medium

The internal filter in position

Left: Dirty water is drawn into the filter at the base, so be sure to position the unit well above the substrate. Once cleaned, the water returns to the tank via the nozzle at the top, which must face diagonally outwards.

separate compartments inside the filter body. The incoming dirty water passes through the mechanical filter medium first – usually open cell foam or ceramic hollow cylinders – to remove solid waste. The water then passes through the biological medium, now often a 'high-tech' sintered glass media which has been baked to a high temperature to create millions of microscopic cracks that are the ideal home for colonies of 'cleansing' bacteria. The final stage is usually shared between a specific chemical medium and/or a wad of fine filter wool to trap any tiny particles of solid debris remaining in the water before it is pumped back to the aquarium.

The flow rate from an external filter is governed by how far it is below the water surface and how long the hoses are. For the best flow, place the filter as close to the water level as possible – typically in the cabinet underneath the aquarium – and minimise the hose lengths. When selecting a unit, ensure that it has good-quality isolation taps, as these will be the only thing preventing a flooded room when you disconnect the filter for maintenance. Also check that the seal between the motor head and the body – usually a rubber 'O' ring – is easy to locate and stays in place when you re-assemble the unit. If this seal is not located properly, the filter will leak.

An external power filter

Dirty water enters the filter by gravity, is cleaned and then pumped back to the aquarium.

Internal canister houses the different filter media arranged in layers.

Biological medium (ceramic cylinders) provides a surface on which beneficial bacteria can multiply.

Filter foam traps large pieces of debris and also acts as a biological medium .

Activated carbon removes toxic substances.

Filter floss prevents any fine particles becoming trapped in the impeller.

Above: Arrange the filter media in the canister, with the coarse filter foam at the bottom, followed by a layer of biological medium, then the carbon and finally a wad of filter wool.

23

DAY ONE

Undergravel filters

Once the most popular choice for home aquariums, undergravel filters have been superseded by internal and external power units. The undergravel system uses a filter plate on the base of the aquarium and this is covered with a layer of substrate that acts as both the mechanical and biological filter medium. Water is drawn up a tube connected to a void beneath the plate, either by rising bubbles from an airpump or more efficiently by an electric powerhead, and returns to the aquarium. This upward flow sets up a circulation so that aquarium water is sucked downwards through the substrate to complete the circuit.

The main drawback of this system is that solid debris is sucked into the substrate and under the filter plate, where it becomes almost impossible to remove and can build up to harbour harmful bacteria and fungus. The second drawback is that it is impossible to grow live plants successfully within such a system. The water movement through the substrate is too severe for the delicate roots, and they simply die back. An internal or external power filter is a much better choice for aquarium filtration.

Below: This small, air-powered filter contains biological, chemical and mechanical media, making it ideal for small aquariums, breeding tanks or hospital/quarantine setups.

Air-powered filters

These compact internal units draw water through a small container of filter medium or a block of foam and are powered by an upward stream of bubbles supplied from an airpump. They are ideal for a quarantine or hospital tank (see page 118), where their action provides aeration as well as filtration for new or convalescing fish. They are not suitable for a main aquarium and certainly not where carbon dioxide dosing is running to boost plant growth. The extra aeration will 'gas off' the carbon dioxide before the plants have the chance to use it. (See page 26 for more details of carbon dioxide fertilisation systems.)

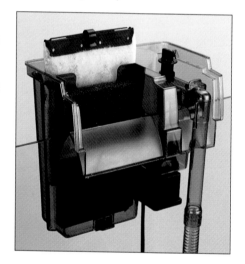

Above: A hang-on power filter on the side of an aquarium, shown with the cover removed and the carbon-filled floss bag pulled up into view.

Hang-on power filters

A hang-on power filter works in the same way as an external power filter, except that it hangs on the edge of the aquarium glass. Water is drawn up into the unit by an impeller, passes through various types of filter medium and then cascades back into the aquarium.

Types of filter medium

Mechanical filter media range from various grades of foam to very fine filter floss. Biological filtration media include

Above: Air-powered sponge filters are safe to use in fry-rearing tanks. Not only do they trap dirt particles, they are also colonised by micro-organisms that provide live food for young fish.

sintered glass and ceramic cylinders. Chemical media are used as short-term solutions to water quality problems.

Open cell foam is used in both internal and external power filters. It provides a huge internal surface area on which beneficial bacteria can grow. It also functions as an excellent mechanical filter medium but needs regular cleaning to ensure that the core does not become clogged with debris. With time, foam will

Below: Filter foam doubles as a means of mechanical and biological aquarium filtration.

shrink and lose its shape and will need replacing. It is a good idea to replace only half the foam at a time to ensure that you do not lose all your filtration bacteria at once. Test the water for a few days after each change of filter foam, as the initial reduction in bacterial population may allow the water quality to deteriorate.

Ceramic cylinders form an ideal first-stage mechanical filter medium in an external power filter. Suspended debris is readily trapped between and inside the rough-textured pieces. Like all filter media, with time they will also carry a population of filtration bacteria; maintain these by cleaning the medium in old aquarium water. Unlike foam, ceramic cylinders and similar materials are relatively easy to clean and never need replacing.

Sintered glass is a high-capacity biological filter medium manufactured at extremely high temperatures to create microscopic cracks in the surface of the fragments. This massive surface area not only supports aerobic nitrifying bacteria (that break down ammonia and nitrites into much less toxic nitrates), but also anaerobic reduction bacteria that break down nitrates into completely harmless nitrogen gas. Use sintered glass exclusively as a biological medium and provide excellent mechanical filtration for the water reaching it. Any prolonged buildup of solid debris will render the microscopic cracks useless as sites for efficient biological filtration.

Gravel was once widely used as a filter medium – principally in undergravel filters – but has been surpassed by more efficient materials. Large quantities are needed for it to be effective biological filter and it is difficult to maintain.

Porous ceramic material ('Alfagrog') has a high surface area for its particle size and can be an excellent alternative to sintered glass. It is often used as a planting substrate (as we have

Activated carbon *blended from a variety of sources is the most effective for aquarium use.*

Filter wool *removes fine particles and 'polishes' the water before it is finally returned to the tank.*

Sintered glass cylinders *provide a huge surface area for beneficial bacteria to colonise.*

Alfagrog *is a porous, inert, ceramic material that supports nitrifying bacteria in the filter system.*

Gravel *plays a part in mechanical and biological filtration, but is not as efficient as other media.*

Chemical filter media *have specific purposes, such as removing nitrates or phosphates.*

in the featured aquarium), where it will also support a healthy and useful population of nitrifying bacteria.

Activated carbon has been used as a filter medium for many years but only recently have we understood how it works. Activated carbons from different sources, such as bone, coconut or coal, have different characteristics and perform certain aquarium roles better than others. The most efficient option, therefore, is a blend of carbons from various sources. The active sites on the carbon adsorb (take up on the surface) impurities from the water and so carbon is useful for

removing such substances as dyes, leftover medications and organic wastes. Change your carbon every month to six weeks and always rinse it under the tap before use to remove any fine dust.

Chemical filter media are usually specific ion-exchange resins that pick up a particular chemical in the water and swap it for a harmless salt. These media are very fast acting and are mostly used to remove nitrates or phosphates. Always use them along with a test kit to monitor the levels of the chemical you are trying to eliminate. Remove chemical media before treating the tank with medications.

Filter wool is ideal as the last stage of filtration to remove fine particles and add a final 'polish' to the water before it returns to the aquarium. Use it either loose or in pad form and check it regularly, as once it becomes clogged it will let very little water through. It is difficult to clean without destroying the structure of the material. The best option is simply to use a new batch of this inexpensive filter medium each time.

Fitting a heater thermostat
Once the filter is in place, the heater thermostat can be fitted. Before you do this, check the temperature setting – most

A carbon dioxide generator

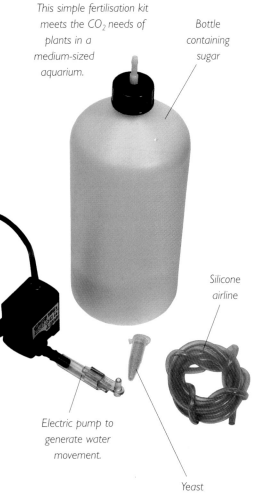

This simple fertilisation kit meets the CO_2 needs of plants in a medium-sized aquarium.

Bottle containing sugar

Silicone airline

Electric pump to generate water movement.

Yeast

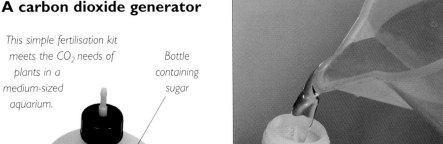

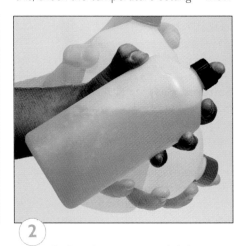

1 *Add warm water to the plastic bottle to dissolve the sugar.*

2 *Replace the screw cap and shake the bottle well.*

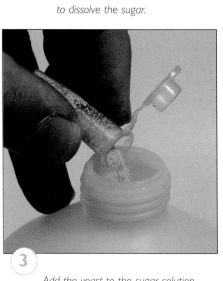

3 *Add the yeast to the sugar solution to set up the fermentation process.*

4 *Connect the silicone airline to the special cap of the reactor bottle.*

Introducing CO₂ gas

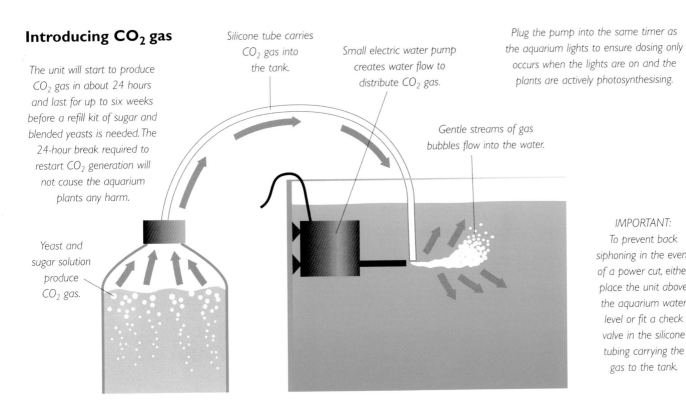

The unit will start to produce CO₂ gas in about 24 hours and last for up to six weeks before a refill kit of sugar and blended yeasts is needed. The 24-hour break required to restart CO₂ generation will not cause the aquarium plants any harm.

Silicone tube carries CO₂ gas into the tank.

Small electric water pump creates water flow to distribute CO₂ gas.

Plug the pump into the same timer as the aquarium lights to ensure dosing only occurs when the lights are on and the plants are actively photosynthesising.

Gentle streams of gas bubbles flow into the water.

Yeast and sugar solution produce CO₂ gas.

IMPORTANT: To prevent back siphoning in the event of a power cut, either place the unit above the aquarium water level or fit a check valve in the silicone tubing carrying the gas to the tank.

heater thermostats are supplied preset at 27°C (80°F). To alter the temperature, simply turn the adjuster knob at the top of the unit. Make sure the model is fully submersible and fasten it to the inside of the back glass using the plastic suckers supplied. Place it at an angle of 45° so that the heat dissipates away over a large area. (If you position it vertically, the thermostat near the top may be affected immediately by heat rising from the heater element directly below.) Important note: Do not connect the filter and heater to the electricity supply until the aquarium is full of water.

Carbon dioxide fertilisation

The final piece of equipment to be added inside the aquarium is a small, low-pressure water pump for dosing and circulating carbon dioxide to promote healthy plant growth. For plants to flourish in the aquarium, they need sufficient levels of dissolved carbon dioxide gas (CO₂) to use in photosynthesis to make glucose for 'energy' and growth. In a natural aquatic habitat, the respiration of other living organisms (including bacteria) produces enough carbon dioxide gas as a 'waste product' to sustain photosynthesis. In the enclosed environment of an aquarium, however, there is not enough respiration taking place to maintain sufficiently high CO₂ levels for optimum photosynthetic

activity. Adding carbon dioxide gas directly to the water makes up for this shortfall.

Where do we get the supply of CO₂ gas? For small to medium-sized aquariums up to 150 litres (33 gallons) in capacity, a simple biological generator system can be used. This consists of a plastic bottle in which a sugar solution mixed with a blend of yeasts sets up a fermentation process that produces a steady stream of CO₂ bubbles. The gas accumulates in the top of the bottle and passes to the aquarium along a silicone airline that opens close to the outlet of the water pump. Water flowing from the pump nozzle draws CO₂ gas from the end of the airline by a venturi effect and distributes it in a stream of tiny bubbles.

The key feature of this system is that it operates at low pressure and the bubbles produced are really small. This makes for an effective and highly controllable system. Adding carbon dioxide to the aquarium must be done carefully: too little, and the plants will not benefit; too much, and the acidic properties of carbon dioxide will change the water chemistry, making it

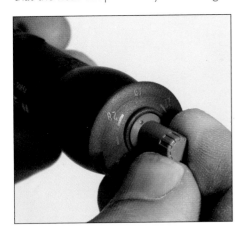

Above: Most heater thermostats can be set to maintain a temperature between 18 and 32°C (64-90°F). They are generally preset to 27°C (80°F), but are easy to adjust by turning a knob.

27

unsuitable, or even lethal, for fish. Simply directing a high-pressure stream of CO_2 gas into the aquarium for 24 hours a day would have disastrous results.

The other crucial influencing factor is the presence of light. Plants need light to carry out photosynthesis and so there is no point in supplying CO_2 gas at night. The solution here is to connect the electricity supply to the water pump into the timer that also controls the aquarium lights. When the lights turn on, so does

the pump. During the night, when the pump is off, any CO_2 gas bubbles escaping from the airline float up to the surface and burst 'harmlessly' into the atmosphere above the aquarium.

A pressurised CO_2 system

For larger aquariums, the best option is to use a carbon dioxide dosing system based on bottles of pressurised gas. This is more expensive but is able to supply the higher volumes of carbon dioxide required. Such

a setup can also support vigorous plant growth for a few months at a time before the gas bottles need to be refilled, which can be done at your local aquarium shop.

As these units are supplying CO_2 gas under pressure, there is no need for a pump in the aquarium to distribute the bubbles. Various types of diffusers are available, such as the bubble counter used in the system featured opposite. The main consideration is that either the bubbles are very small, which gives a large surface

Right: Place the pump at the opposite end of the tank from the filter to boost the circulation of water around the aquarium. Fix it no deeper than 15cm (6in) below the eventual water surface; any deeper and it would reduce the rate of carbon dioxide diffusion.

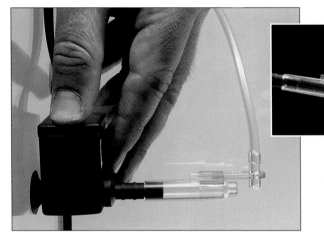

Left: CO_2 bubbles are released from the tube. This picture shows the principle, although in reality the stream should be much gentler.

Below: Although the filter and heater are vital pieces of aquarium equipment, from an aesthetic point of view it is better if they are hidden from view. Clever planting later on will conceal them, yet leave them accessible for routine maintenance.

area per bubble for CO_2 to dissolve into the water, or that there is a long contact time between the gas and the water to ensure that the required volume of CO_2 diffuses into the aquarium. To check that the correct dosage of CO_2 is going into a large planted aquarium, you can install a permanent CO_2 level indicator. This fits on the inside glass and changes colour to indicate too much, too little or the correct amount of CO_2 in the aquarium.

As in the smaller scale 'biological' system we are using in our featured aquarium, a pressurised setup must only dose the aquarium with carbon dioxide when the lights are on. This is usually achieved by fitting a solenoid valve in the gas line that is connected to the lighting timer. When the timer turns the lights on, the solenoid valve opens to release the supply of CO_2 gas into the aquarium.

With a pressurised CO_2 dosage system, it is good idea to have the returning water flow from your aquarium filter passing through or near the diffuser to ensure thorough mixing of dissolved CO_2 around the aquarium.

CO_2 cylinder fertilisation

Cylinder systems can be connected to a light timer so that gas is only released when the lights are switched on. Plants have no use for CO_2 at night and an excess at night can harm the aquarium.

This valve closes when the lights are off, preventing the release of CO_2 gas.

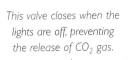

The aquarium lights provide an energy source for photosynthesis.

The cylinder contains compressed CO_2 gas, which is released at a controlled rate via a regulator. The two dials indicate the release rate and the pressure of the gas in the cylinder, which reflects the amount of gas remaining.

The bubble counter allows tiny CO_2 bubbles to travel slowly upwards, allowing maximum time for the gas to diffuse into the water.

Left: *Carbon dioxide gas enters at the bottom of this diffuser, or bubble counter. After a running-in period of about 48 hours, the bubbles stabilise and become smaller as they rise and release CO2 gas into the aquarium water.*

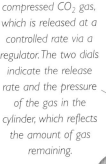

Connectors and valves are standard fit and suitable for all systems.

Left: *Cylinders containing compressed CO_2 are ideal for larger aquariums and long-term CO_2 fertilisation. The gas produced is sent to a bubble counter, where it is kept in contact with the water for an extended time.*

29

PREPARING AND ADDING WOOD

Bogwood and bamboo are popular items of aquarium decor that simulate natural features from various different habitats around the world. The safest option is to buy bogwood from an aquarium shop, as you can be sure that it will be suitable for underwater use. Wood from any other source is best avoided, as it will quickly rot and may release chemicals that could be detrimental to the health of your fish and plants. If you have any doubts about the source of any wood, you should seal it with clear plastic paint normally used for sealing concrete ponds. This product is safe to use with fish and plants. It not only prevents any harmful substances leaching into the aquarium, but also stops the wood rotting. Several coats may be required to achieve a good seal. Make sure the wood is completely dry before you paint it. The sealing process may give it a slightly shiny appearance,

but this will soften over time as the bogwood takes on a natural look.

Bogwood is available in many sizes, shapes and colours, and a good aquarium shop will offer a wide selection of pieces. Most of the differences in finish are created by the treatment the wood receives at source. Most pieces will be long-dead tree roots that once grew in boggy conditions, hence the name. As a guide, look for pieces that resemble tree roots and stumps, as these will look more natural in the display. Some pieces can be held in position using aquarium suckers to give the impression of a larger piece of wood that extends beyond the boundaries of your display. Bogwood can also be used to hide unsightly pieces of equipment, such as an internal filter.

When using bogwood always following the preparation guidelines shown below. Once the bogwood has been soaked –

Above: Brush the bogwood to remove dirt and debris. Wet it to remove stubborn marks and use a smaller brush to get into the narrow crevices.

and this may take up to two weeks – place it in position in the aquarium. Preparing the wood usually ensures that it sinks when you add water to the aquarium, but if it floats, there are a couple of steps you can take to anchor

Below: Place the bogwood into clean water in a bucket deep enough for the wood to remain submerged. Weight it down if it floats.

After several days you will notice that the water has started to discolour. Leave the bogwood submerged in the bucket.

When the water looks like strong tea, replace it with fresh tapwater. Repeat the process as often as necessary to remove most of the tannins.

It may take up to two weeks and several changes of water before the water remains clear. At this point, you can place the clean, waterlogged bogwood into the aquarium.

it securely. One method is to tie the wood with thin nylon fishing line to a stone that will be hidden from view. Alternatively, silicone the wood to a piece of glass and then bury the glass in the aquarium gravel. Any obvious joins can be easily disguised with planting.

With the bogwood in place, stand back and take a look at it, ideally from the angle at which you will normally view the aquarium. This will enable you to see if it looks natural and whether you are happy with its position in the overall display.

Preparing bogwood

Always ask if the bogwood you are buying has been presoaked; in most cases, it will not have been and you will have to treat it before use. The reason for soaking the wood is that it contains large quantities of natural humic acids and tannins. If you simply place the wood into the aquarium, these naturally occurring chemicals will leach out into the water

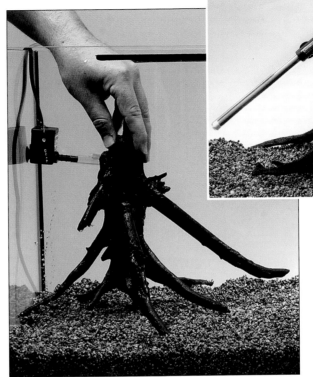

Left and above: Bed the bogwood into the substrate. In this setup, there is a piece on each side. It can be used to hide aquarium equipment, but make sure you can still access heaters and filters. According to their shape, arrange the pieces so that they resemble tree roots and stumps or fallen branches.

Types of wood

Bogwood is available in many shapes and sizes. If a piece is too large for your tank or has a sharp corner, break it with your hands for a natural finish.

Twisted roots are thinner pieces of wood that can represent overhanging branches or roots growing in a riverbank.

Mopani wood is precleaned and has a different texture on each side.

Cork bark is useful for hiding tank equipment, but will need anchoring to prevent it floating.

and stain it a tea brown colour. The chemicals are not harmful, but most fishkeepers prefer to keep the water as clear as possible.

To prepare bogwood, first brush it thoroughly to remove any dust and dirt. Then place it in a bucket of water, making sure it is totally submerged. Leave it for several days so that the tannins are released. If the wood floats, simply place a heavy object on top to keep it submerged. Once waterlogged, it will not float when you put it in the aquarium. The water in the bucket will gradually discolour until it resembles strong, dark tea. At this point, discard it and replace it with a fresh supply. Repeat the process until the water remains clear for several days. The wood is now ready to use.

Even though you have presoaked the bogwood, tannins will slowly continue to leach into the aquarium and these will gradually cause the pH level of the water to drop. This is something you scan monitor with the help of a pH test kit (see page 96), but a maintenance routine that includes regular water changes should prevent any problems.

Bamboo

The sealing process described for wood on page 30 is also suitable for two other types of wood that you can use in the aquarium, namely bamboo and cork bark. With its unique shape and texture, bamboo is a striking decor material and can provide excellent cover for many aquarium fish. It is available from aquatic and garden centres in various diameters. To anchor the bamboo, you may also need to attach it to a piece of glass. Use aquarium silicone sealant for this job as it is strong, chemically inert when dry and therefore harmless to fish and plants.

Right: The shape and rough texture of this 'Jati' wood, a type of bogwood, are reminiscent of fallen branches. The natural 'cave' formed by the wood will be a welcome retreat for the fishes. Bed down all decor securely to avoid the risk of collapse and resulting damage to the tank and its inhabitants.

Cork bark

The same sealing and fixing processes can also be applied to cork bark. Its chunky texture can be almost rocklike and some aquarium plants, such as *Anubias*, will readily grow on it. Cork bark is available in various lengths; if you need to shorten

Above: Left untreated, bamboo will rot in the aquarium, releasing organic elements that may cause algal or fungal blooms. Allow it to dry out completely before painting it with a sealant suitable for aquarium use.

a piece, break it with your hands rather than cut it as this will result in a more natural finish. Cork bark is extremely buoyant and must be attached to a piece of glass with aquarium sealant to keep it submerged, as shown opposite. Anchor the glass securely below the substrate.

Below: Thick bamboo canes, cut to different lengths, weighted down with a cobble inside and positioned amid dense planting, become distinctive items of decor in this display.

Aquascaping

If you are looking for inspiration to create an exciting underwater landscape (or 'aquascape') in your aquarium, then start by looking at tank displays in shops or in other hobbyists' homes. If you want to create a themed display, books and magazines feature plenty of location photography of the natural habitats in which aquarium fish and plants live. But whatever style of aquarium you choose, remember that most successful displays are made up of just a few items of hard decor; a display containing examples of every type of rock and wood does not look realistic. Having chosen a few materials, concentrate on the shapes and sizes of the pieces you need for the display.

Before placing any decor in the aquarium it is a good idea to sketch out an overhead plan showing where all the elements are to go. Think of the process as creating a frame in which the fish will be the living picture.

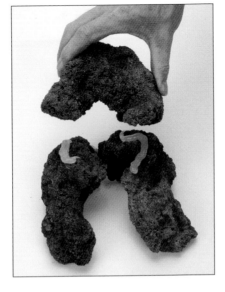

Above: To make a lava rock cave, select a large piece of rock for the base and apply sealant as necessary. Firmly press on the 'roof', fill gaps with sealant and leave to dry.

Bear in mind that plants will require sufficient room to grow and flourish. With all the filter and heating equipment installed in the aquarium, you can see which items you need to hide and how much space you have to play with. If you prefer, set up the wood, rocks and other items of decor outside the aquarium in a 'trial run' to see how they all fit together. However, if you are confident, you can place them directly into the tank. The best way is to start at the back and work forwards, beginning with the largest piece of decor. Always take great care when installing rocks and wood not to damage either the aquarium or any of the equipment.

Continue to build up the display from back to front, leaving room for the plants. For most rectangular aquariums, a 'half bowl' shape is ideal, with the tall items at the sides and rear of the display and lower ones towards the front and centre. This will frame the picture and also leaves an area of open swimming space where you can most easily see the fish.

Guidelines to help you create a realistic display:

- Rock strata should run in the same direction, ideally horizontally to reflect most natural formations.
- Wood, especially buoyant pieces, must be securely anchored.
- Root-shaped wood positioned vertically gives the impression that it is part of a tree above the aquarium.
- Smaller pieces of rock or wood can divide or define plant-growing areas.
- Use background decor to hide aquarium equipment.
- Do not rush things. If possible, set up what you think is a good display, leave it overnight and see if you still like it in the morning – well before adding the water.
- Empty space is not necessarily a bad thing; it represents growing room for plants and a swimming area for the fish.

Above: Siliconing small pieces of bark to a piece of glass prevents them floating. Placed on the aquarium floor, they appear to make a long, but fragmented line.

A basic plan

Mark in where all the equipment is to go. This is the water pump connected to the CO_2 fertiliser unit.

Sketch in the intended location of the rocks you propose to use.

Include the filter and heater-thermostat in a position where they can be easily accessed for routine maintenance.

Mark in the location of the bogwood pieces. Leave an open swimming space for the fish.

ADDING ROCKS

Aquarium shops carry a selection of rocks, but not all of them are suitable for the freshwater tropical community display we are creating here. As with the substrate, the requirement is for inert materials that will not affect the water chemistry. Ideal rocks include basalt, flint, slate, sandstone, quartz and lava rock. Unsuitable materials include limestone, marble and chalk. Avoid these calcareous rocks, as they will continually raise the pH of the water in the aquarium by releasing calcium salts.

Do not collect rocks from the wild, as you may not be able to identify them and, again, they could alter the water chemistry in the aquarium (see page 36). Here we look at a range of rocks that are suitable.

Cobbles and pebbles are an excellent addition to the display. They have a natural riverbed appearance and are ideal for creating a textural difference in the foreground. They can also be used as a dividing line between two vigorous foreground plants, such as *Glossostigma elatinoides* and *Echinodorus tenellus*.

Lava rock ('Alfagrog') can be used in large pieces to create a completely different look. With this material you can create a backdrop that is ideal for growing plants such as Java Fern, *Anubias barteri*, *A. nana* and *Monosolenium tenerum*. The roots of these plants have no problem anchoring themselves securely onto the open, porous structure of the lava rock. It is always worth running your hand carefully over the lava rock surface before placing it in the aquarium to check for very sharp edges that could injure the fish. They will be difficult to remove later on, so knock them off now with a hammer.

Coal (if you can find it) is an unusual addition choice of decor, but can look very effective. Wash it thoroughly before use – a messy business that may persuade you to use another material.

Westmorland rock is an attractive addition to the aquascaping armoury, although not the cheapest rock to buy, and not always available. The red-brown colour and striated markings make it a distinctive alternative for use in a brackish aquarium display.

Slate is commonly available and can look stunning. Use the flat, purple-grey pieces to create some interesting aquarium features, such as a natural cave. Place a piece of slate onto the substrate, leaving just a small gap underneath and any fish with a slight burrowing instinct will shape the substrate to their own requirements.

Types of rock

Granite has an unusual sparkling texture and will appear more natural with time.

Slate is available in a range of shapes and sizes. Position this heavy rock with care.

Rounded boulders have a great impact in aquarium displays and make good grazing spots for algae-eaters.

Westmorland rock has an attractive reddish colour and striking markings.

Chippings are available for many rock types. These are slate pieces.

Broken pieces of lava rock create an unusual substrate.

Washed coal is a good choice for creating a darker display.

Small rounded pebbles are ideal for creating a riverbed.

Granite, flint and sandstone are all inert rocks for use in the aquarium. You can combine smaller chippings with large pieces for extra effect. If your aquarium shop only has a limited selection of pieces, consider a visit to a local landscape centre for more choice. Apart from an aquarium shop, this will be the only place where you can be sure of what you are buying.

Artificial decor

Finally, do not overlook the range of artificial decor available. Modern products look very realistic, as the moulds used are often taken from real pieces of wood or rock. The paint finishes are excellent and you can build up a very realistic backdrop using these pieces. Many have also been developed to hide specific items of equipment inside the aquarium.

Above: Individually, these cobbles, pebbles and stones make good aquarium decor, but used together, they have greater impact.

Below: Combining artificial pieces of decor with genuine rocks and plants help them to appear more authentic. Here the large and small rocks and pebbles are real, but the 'wood' is synthetic.

This fake bark is a good shape for hiding internal filters.

Artificial wood looks natural and does not alter the colour of the water.

Synthetic rock is inert and safe for any aquarium.

Fake wood is available in manageable sizes.

Choosing and using rocks

When selecting rock, take time to examine several pieces and arrange them to see how they look together. If the rock has visible strata, the golden rule is that all the lines should run in the same direction. Arranged properly in the aquarium, they will have the appearance of a natural river- or lake-bed. As with bogwood, rocks can be positioned to define different areas of the aquarium and to form a natural divide between plants.

If you are unsure about the suitability of a rock, then a simple vinegar test will put your mind at ease. Simply place a few drops of malt vinegar onto a clean area of the rock surface. If the vinegar begins to fizz, it is the result of a chemical reaction between the acidic vinegar and the calcareous material of the rock.

Preparing rocks

Before using any rocks in the aquarium, scrub them to remove dust and debris. Clean, wet rock shows its true colours under aquarium lighting. Dirty rock looks out of place underwater; in the natural world, it would have had many years of washing. Dust from rocks will also cloud the water, which spoils the overall effect.

Testing rocks for suitability

To test if a rock is likely to alter the water chemistry, pour on some acidic substance such as vinegar. If the rock contains any calcareous substances, it will begin to 'fizz' gently. If there is no fizzing it should be safe to use in the aquarium. You may need to examine the rock carefully for the bubbles that appear when an acid is added to an alkaline rock.

Below: Rocks are not only dirty and dusty, they may also harbour mosses and lichens that could foul the water. Scrub them thoroughly before adding them to the tank.

You may prefer to wear rubber gloves while you scrub the rocks.

Low-pressure C0$_2$ fertilisation system

A small brush with stiff bristles will penetrate all the crevices.

Wash the rocks in clean water. Working over a bucket minimises the risk of splashing water on the floor if you are working indoors.

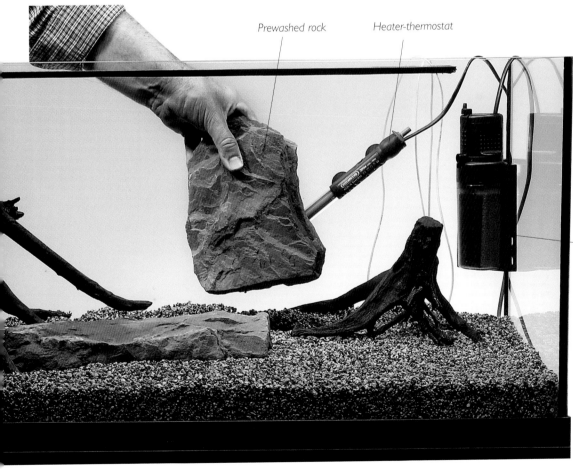

Prewashed rock

Heater-thermostat

Internal
power filter

Above: Decide where the rocks are to go. If you are following a plan on paper, refer to it as you work, making adjustments as necessary. When you are satisfied, bed the rock into place.

Left: Tall rocks are useful for hiding unsightly equipment. Lower heavy pieces carefully into the tank to avoid damaging the glass. Make sure they are stable and secure.

TREATING AND ADDING WATER

Only when you have installed all the aquarium equipment and are completely happy with the layout of the rocks, bogwood and other decor, is it time to fill the tank with water. Do not add plants at this point, as most require the support of the water to stay upright and in any case, the filling process would dislodge them.

Conditioning the water

As we discuss on page 92, tapwater is ideal for the aquarium, providing it is first treated with a proprietary water conditioner to remove chemicals that would otherwise harm the plants and fish. A product that also contains beneficial bacteria will help to trigger biological cycles in the aquarium. Following the manufacturer's instructions, simply add the whole of the required amount of liquid water conditioner to the first bucketful of tapwater before filling the aquarium.

Final checks

Before you start to fill the aquarium, make sure that it has not been knocked slightly out of place during the first phase of setting up. A tank that overhangs the base, even by a small amount, is both unsightly and dangerous, and the situation will be impossible to rectify when the tank is full of water. It is also worthwhile installing any tight-fitting hood surrounds before the aquarium is half full. The water may cause the glass to bow slightly, especially in deep aquariums, and hoods may prove very difficult if not impossible to fit later on.

Below: *Pour the conditioned water carefully from the bucket onto a saucer, flat stone or even a plastic bag to prevent the water splashing. If you prefer, start filling the tank with a plastic jug and progress to a bucket when the water level has started to rise.*

Measuring the tank volume

Filling the aquarium with a bucket of a known capacity enables you to measure the volume of water in the aquarium accurately. Although it is easy to calculate the volume of a rectangular aquarium, this does not take into account the displacement of the gravel, decor and equipment. Using a bucket, you can measure the volume of water precisely. When the aquarium is roughly two-thirds full, record the water volume so far. It is important to know exactly how much water the aquarium holds, because at some future date you may have to treat sick fish with medication diluted according to the volume of the aquarium. Accurate dosage is vital; underdosing may not kill the disease organism you are treating and overdosing can quite easily kill your fish.

Above: *Using a tapwater conditioner is a quick and reliable method of neutralising the chemicals added to make the water suitable for human consumption. It is then safe to use in the aquarium.*

Above: *This model of aquarium is supplied with a tight-fitting surround. It is a good idea to install it now, in case the water causes the glass to bow slightly when the tank is full.*

Filling the tank to this level minimises the risk of water overflowing when you introduce the plants.

Below: *Continue filling the tank until it is two-thirds to threequarters full. However careful you are not to disturb the substrate, the water is bound to look cloudy at first.*

Filling the aquarium

As you fill the tank, it is important not to disturb the carefully washed gravel and layer of planting substrate. To avoid too much splashing caused by a strong flow of water from a bucket, place a plate or a thick plastic bag onto the gravel and slowly pour the water onto the plate or bag. This will help to spread out the water and minimise any disturbance.

Initially, you should only fill the aquarium to two-thirds of its full capacity. This will leave plenty of room for water movement and displacement when you put your arms in the tank to plant it up. The water you have put in will be sufficient to support the plants once they are in position.

PREPARING TO PLANT UP THE AQUARIUM

Now that the bare bones of the aquarium are in place, the enjoyable task of completing the aquascaping can continue. Overnight, the filter will have cleared any haziness from the water and the heater will have warmed the water to its preset temperature. The result is a clean and comfortable environment in which to build up the display.

Before proceeding with any planting, it is worth returning to your basic plan of the aquarium and sketching in where the plants are to go. Keep the plants in their transportation bags until you are ready to use them. In this way, they will remain in a moist environment and will not suffer too much from drying out, which can quickly happen in a warm house. Each plant can then be dealt with in turn, without being exposed to the drying air for too long. Start by adding the background plants and working forwards to complete the display.

Here are a few final tips before you begin planting. Wear short sleeves when undertaking any tasks that involve putting your hands in the tank. Keep a towel within easy reach, as your arms will be constantly in and out of the aquarium, which can lead to plenty of drips and a soggy carpet. If you have sensitive skin, consider wearing rubber gloves to prevent any problems arising from prolonged contact with the water. Finally, keep a waste bucket close by so that you can easily dispose of packaging materials, plant baskets and unwanted leaves.

By the end of the day you should have an immature but complete planting display that is ready to flourish under your guidance.

After 24 hours, the water will have settled and cleared and the aquarium is ready to be planted up. Later in this section you will find a selection of plants for all areas of the display.

CHOOSING PLANTS

Adding plants to your aquarium is one of the most creative aspects of building up your 'living picture'. They form the canvas on which your fish will display. The choice of aquatic plants is extensive, with an array of shapes and sizes every bit as diverse as in terrestrial plants.

Where to start?

Faced with such a choice of plants, your first consideration must be the size of your aquarium and the type of display you want to create. Although you may love a particular plant, there is no point in including it if it will outgrow your aquarium in six months time. Some aquariums look stunning with minimal planting in a 'geological display' dominated by rocks and bogwood, while others are totally plant filled, with no other decor visible amongst the foliage. Both displays can be equally effective, even though they are completely different. Or you may want to recreate a natural habitat, such as an Asian stream, using only the plants and fish that are found together in the wild.

Making a plan

On page 33 we discussed sketching out an overhead plan of the aquarium showing where all the major items of equipment and hard decor should go. You can take this idea a step further by also making a planting plan in the same way that you would design garden beds. With the fixed 'hard landscaping' in place, you can see how much room is left for plants and where the planting areas occur.

You can divide the plan to create background, midground and foreground areas. These are not intended to be strict lines within which to plant, but they will be useful as a guide to help you design the display. To achieve a natural half-bowl, tiered effect, envisage your plan from the front to the back, with the shortest plants in the centre foreground and the tallest ones at the back. Also consider having a tall plant at each of the front corners. The finished effect should look natural, with plants overlapping and not arranged like a regimented flowerbed. Be flexible with your planting scheme and make sure you leave enough room for each plant to mature and grow. If the display is crammed full of plants from the outset, they will simply outgrow the space and the weaker ones will die. Turn to the plant section beginning on page 54 to find out how much room each plant will need.

Once you have this information, you can begin to choose suitable plants for each of the three main planting areas.

Buying plants

You do not have to buy all your plants at the same time, but the aquarium display will look more complete if you can finish the planting all at once.

Most good aquatic retailers have a wide range of tropical aquarium plants, which should be displayed in easy-to-view, clean tanks. Plants should be labelled with details of their mature height and spread, price, and ideally, a guide to the best planting position in the aquarium: background, midground, etc. If a certain

Right: When you buy new plants, each one should be packed in a plastic bag with plenty of air inside to prevent crushing during transit. The bags need not contain water; sealing them is enough to retain moisture.

A planting plan for your aquarium

Background

Midground

Foreground

Water pump for carbon dioxide system.

Mark in the position of any pieces of bogwood.

Heater thermostat

Internal filter

The best effects are created by using plants in groups of three or more of the same species, rather than randomly placed individuals.

Sketch the position of the main rocks.

Use simple shapes and colours to represent plants in the main planting zones.

Choosing healthy plants

Good-quality plants will repay you by establishing quickly and making strong, healthy new growth. Examine plants carefully and reject poor specimens.

Left: *Leaves with no holes, plus roots and small plantlets growing through the base of the pot, are evidence of a healthy plant.*

Right: *Avoid plants with brown or damaged leaves. They will rot away in the aquarium. Yellow patches, particularly towards the leaf edges, indicate a lack of nutrients.*

Above: *Buying plants directly from a retailer enables you to pick out the species you need and examine individual specimens.. However, the range of plants may be limited and you may have to visit several suppliers to locate all your chosen plants.*

species has any special care requirements, these should also be made clear.

It is vital to start with strong, healthy plants. Weak, sickly specimens are unlikely to recover fully, will probably die and generally look unsightly. Always choose plants with evidence of strong new growth and avoid those with many yellowing, blotched or damaged leaves.

Most plants are sold either as potted specimens or in bunches of individual stems secured in a foam-lined lead strip. The lead ensures that they sink and are

Above: *It is vital that mail order plants are well protected during their journey. Here, they are packed in separate bags, placed in a polystyrene box and wrapped in several layers of bubblewrap. Many retailers won't guarantee plants shipped to areas with cold winters.*

Left: *Unpack each plant but leave it in its bag until you are ready to use it. Check the plants against your order. It may be worth ordering more than you need in case of any losses in transit.*

displayed effectively in the aquarium shop. Carefully examine bunched plants and reject any that have been damaged by crushing, as they will take a long time to recover.

In potted plants, you can often see new growth appearing through the sides of the containers – a sign that the plants are in excellent condition. If the roots are visible, they should be white and healthy. Inspect the roots of any plants growing in substrate before committing to buy them.

When you have chosen your plants, the retailer should place them in plastic bags that are then filled with air, which prevents the leaves being crushed during transit. A good retailer will not put all the plants in one bag and you should reject any that appear to have been damaged by rough handling.

Once you reach home, transfer the plants into the aquarium as soon as possible. If your tank has only recently been set up, make sure the water temperature is at least 25°C (77°F), otherwise temperature shock will damage the plants.

Mail order plants

If your local retailer does not stock a wide range of plants or cannot obtain the species you would like, then consider contacting specialist mail order plant suppliers. If possible, ask other hobbyists for recommendations or read the reports in aquatic magazines to check on the quality of the plants on sale. When the plants are delivered, they should be carefully wrapped in plastic bags, surrounded with insulation material and packed in a sturdy box for protection. They are usually despatched by overnight courier, so make sure your aquarium is ready to be planted up before you place an order. It is also worth noting the company's returns policy and procedure should the aquarium plants turn up in poor condition.

Preparing for planting

Good preparation before planting will ensure that plants get a good start in life.

Potted plants The first task is to remove potted plants from their plastic containers. If squeezing the container is not enough to release the plant, cut away the pot and separate it from the root structure. If necessary, cut the pot into pieces to make it easier to remove. Handle the plants with utmost care to avoid damaging delicate stems and roots.

Most potted plants are grown in a rockwool medium that supplies them with nutrients while they are in the

Preparing Echinodorus

This healthy plant is Echinodorus *'Red Special'. The roots are growing strongly through the rockwool medium and out of the base of the pot – always a good sign. Keep any labels that you remove from the pots in a safe place. You may need them for future reference.*

① *Begin by removing the plastic container. Cut the sides of the pot, in several place if necessary, until you can remove it easily and without damaging the plant.*

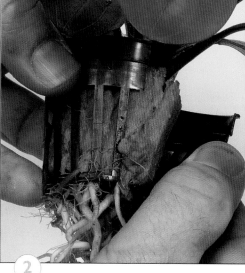

② *Carefully peel away the pot to reveal the rockwool growing medium. Sometimes it is possible to remove the pot simply by squeezing it gently and sliding out the plant.*

nursery. However, once in the aquarium, they will obtain all the nutrients they require from the substrate and a regular programme of fertilisation. Before planting, you should remove as much rockwool from the roots as you can. Peel it away carefully. If some pieces are difficult to reach, leave them in place, as it is not worth damaging the plant just to remove a small piece of rockwool that will eventually be buried in the aquarium.

Once you have removed the pot and the rockwool, examine the plant carefully. Healthy roots should be white and strong. Trim off any with minor damage using a sharp knife or scissors. If you leave them on they will die and decay in the substrate. Remove any leaves that are past their best or have been accidentally damaged in transit. Take off yellowing leaves or any with holes in, again using sharp scissors or a knife for a clean cut.

Bunched plants The lead strips holding bunches of stems in place are not needed in the home aquarium, so undo them carefully and discard them. You could return the strips to the aquarium shop for

Preparing Glossostigma

Glossostigma elatinoides is a tiny foreground plant that only grows to 2-3cm (0.8-1.2in) tall. It requires careful handling.

reuse to avoid dumping them into the environment. Separate out large-leaved plants supplied in bunches and place them into the aquarium in a small group. Long-stemmed, small-leaved aquarium plants look much better if planted relatively close together as a bunch.

Remove as much rockwool as you can without causing damage. If this is too difficult, expose only the base of the roots, leaving the remaining rockwool intact.

Remove as much of the growing medium from the roots as you can, taking care not to damage the roots. This can be quite tricky, so take your time.

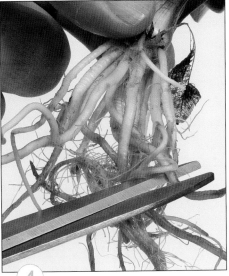

If the roots are too long, trim them with sharp scissors so that you can spread them out when you come to plant them. Roots planted in a bunch may rot.

Check over the plant and remove any old or damaged leaves. Make sure there are no snails or evidence of snails eggs. These appear as blobs of jelly on the leaves.

Planting

Placing plants in the aquarium takes a bit of practice. If you are inexperienced, the easiest way is to use both hands. Hold the base of the plant in one hand, carefully supporting the leaves with your fingers, scoop a hole in the gravel with the other hand and gently push the plant into the hole so that its roots are in the nutrient layer and the crown is just above the finished substrate level. Backfill the

Below: Prepare each plant as described on page 44-45. Working across the back of the tank, begin to put in the plants. This is one of two Echinodorus cordifolius ssp. fluitans that will occupy the back lefthand corner.

hole with the excavated substrate. Avoid disturbing the substrate too much, as you will also move the nutrient layer. Try to work fairly rapidly, as any hole made in an underwater substrate tends to fill up again quite quickly.

If possible, learn to work one-handed. Holding the plant as before, use one or

Right: Moving along to the centre back, the next plant to go in is Echinodorus 'Red Special'. It will develop into a good specimen plant, is not difficult to look after and grows reasonably quickly.

two fingers to make the planting hole and then push in the plant straightaway and backfill with gravel.

When you have decided on the position of the plant in the aquarium, look at it from different angles to see if it has a natural 'front' face. For example, it may be growing unevenly, so that you can see into

When adding plants, be careful not to dislodge items you have already positioned.

Left: *In the right-hand corner, three Echinodorus 'Red Flame' complete the background. Look at the aquarium from the front to make sure you are happy with the position of the plants.*

Below: *Make sure that each plant has its roots in the nutrient layer of the substrate so that it can obtain maximum benefit. Although you must treat the plants gently, firm them into the substrate so that they do not tilt or fall over.*

the crown. For added interest, plant it with the tops of the leaves facing towards the front of the aquarium, instead of leaving the undersides on view.

Background planting

Following the plan you have drawn up, start planting from one corner to the other across the back of the aquarium and then work forwards in the same way to complete the planting.

For the background of this aquarium display, we have chosen three varieties of Amazon swordplant. The first is *Echinodorus cordifolius*, a large, bright green species that will eventually fill the rear lefthand corner and conceal the carbon dioxide pump as it grows. Two plants will begin to form the frame, softening the hard edges of the aquarium and helping the bogwood to blend into the display.

The centre of the background is often a key location for a specimen plant. We are using *Echinodorus* 'Red Special', which has contrasting red-and-green leaves that will create a superb backdrop.

The righthand side of this display is dominated by the upright rock, the heater-thermostat and the internal filter. Although it is important to retain access to the equipment, there is nothing to prevent the development of large-leaved plants that will effectively disguise the hardware when the display is viewed from the front. Three smaller specimens of the stunning Amazon swordplant *Echinodorus* 'Red Flame' are planted in an L-shape around the corner of the tank so they grow up in a group.

Preparing Cryptocoryne

Cryptocoryne wendtii 'Tropica' has attractive, dark, textured leaves. A small group planted together will develop to form a clump in the midground.

1 *Having removed the plant from its pot and teased away the rockwool from the roots, trim away any yellowing or damaged leaves with sharp scissors before planting.*

2 *This cryptocoryne is ready to plant, along with two more, in the midground. Cryptocorynes are popular aquarium plants and widely available. To prevent the risk of cryptocoryne rot, avoid fluctuations in lighting and heating conditions.*

Right: *Supporting the cryptocoryne with three fingers, make a hole in the substrate deep enough to accept the plant roots.*

Midground planting

In this display, the midground will feature one species of plant, namely *Cryptocoryne wendtii* 'Tropica'. Its red-and-green leaves will blend with the Amazon swordplants in the background, but will be perfectly visible, as this species only grows half as tall. The cryptocorynes will form a curtain across two-thirds of the midground, leaving the fish with a swimming area towards the lefthand side near the bogwood and at the rear of the display. Remember to leave enough growing room so that the plants do not become overcrowded. Here, the three *C. w.* 'Tropica' are planted about 10cm (4in) apart. Planting in groups of three or five looks more natural. Of course, there are many more *Cryptocoryne* species – as well as other midground plants – from which to choose (see pages 62-75).

Above: Gently push the plant into the substrate, making sure that the roots are in contact with the nutrient layer. Practise working with one hand; it will prove less disruptive for the existing plants.

Above: With the plant in position, draw in the substrate around the roots to anchor it, just as if you were planting into the ground. Repeat the process with the other two cryptocorynes.

Above: You may find that you need to add some additional substrate around the plants. Drop it gently into the tank. Leave enough space between the plants for them to grow and spread.

Foreground planting

To complete the planting of this display we need some foreground plants, and have chosen four very different species that each make an impact in their own way. The lefthand side will feature *Hygrophila polysperma* 'Rosanervig', a tall plant that reaches the water surface and needs regular clipping back to ensure that it does not overtake the display and block out light to the other foreground plants.

In the mid-foreground there will be two species: the first is *Hemianthus micranthemoides*, a bushy plant with small, fine leaves. It can grow up to 15cm (6in) tall, so it, too, will need clipping to ensure that it spreads along the front of the tank instead of reaching for the light. Two of them planted together will form a clump towards the front.

Below: *Hygrophila polysperma is the choice for the lefthand corner, in front of* Echinodorus cordifolius ssp fluitans. *This end-on view shows the* Hygrophila *being offered up into position.*

Slightly right of centre is a liverwort (*Monosolenium tenerum*), which adds another leaf texture at the front of the display. Like most liverworts, this one does not root into the substrate, but is supplied attached by netting to a rock or purpose-made planting stone. Simply place this on the substrate and the plant will soon cover the entire stone and the netting. The effect is of an anchored plant growing naturally in the substrate.

In time, the front righthand side of the display will be covered by a lush green carpet of *Glossostigma elatinoides*. Three of these low-growing, round-leaved plants will fill the gaps between the specimen plants. We are leaving plenty of space between them for growth, but this invasive species will need controlling.

Together, the four plants should create an interesting foreground, allowing the fish plenty of swimming room, but also providing cover should they require it.

Below: *Make a hole in the substrate and plant the* Hygrophila *in the usual way. Pull up the substrate around the base of the plant. This is the view from the front of the tank.*

Planting Glossostigma elatinoides

To the right of the foreground there are three Glossostigma elatinoides. Given sufficient light, these will make a dense carpet across the substrate, filling in any gaps.

Plant the three Glossostigma, leaving room for them to spread.   In time, the plants will merge together to create a seamless carpet of plants at the front of the aquarium.

Left: *Next to the Hygrophila, the foreground display continues with two plants of Hemianthus micranthemoides that in time will form a bushy clump.*

Right: *In the wild, Monosolenium tenerum forms cushions on stones. For aquarium use, you can buy it already attached to a 'clay' stone that you simply place on the substrate, but can move later on if necessary. As it grows, it will entirely cover the stone.*

THE PLANTED AQUARIUM

Above: *As you plant up the tank, take a look at it from various angles. If the sides are visible, make sure there is some planting of interest there. This view from the righthand side of the tank lends a whole new perspective.*

Right: *The larger species of Echinodorus are very effective for the background of the aquarium, and this 'Red Flame' makes a striking focal point. For the best results, provide a nutrient-rich substrate, good lighting and regular iron fertilisation.*

Hygrophila polysperma 'Rosanervig'

Echinodorus cordifolius ssp. fluitans

Hemianthus micranthemoides

Room for development

Do not be tempted to fill the aquarium with plants straightaway. Allowing the existing plants to spread naturally will always look more natural. If gaps remain, you can always add more plants later on.

Echinodorus 'Red Special'

Cryptocoryne wendtii 'Tropica'

Echinodorus 'Red Flame'

Echinodorus 'Red Flame'

Echinodorus 'Red Flame'

Monosolenium tenerum

Cryptocoryne wendtii 'Tropica'

Glossostigma elatinoides

Cryptocoryne wendtii 'Tropica'

Glossostigma elatinoides

The plan in action

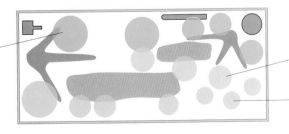

Background

The large Echinodorus *species used here create a strong structural background to the display.*

Midground

Cryptocorynes define the midground area and will soften the rock edges.

Foreground

Low-growing Glossostigma *and* Hemianthus *leave swimming space for fishes.*

53

PLANTS FOR YOUR AQUARIUM

This part of the book features a selection of aquarium plants that will thrive in most conditions. They are grouped according to the main planting zones available within the aquarium display.

The background should be dominated by the tallest species, ideally growing up to the water surface. They can vary in appearance, from the ribbonlike leaves of the *Vallisneria* species to the spade-shaped leaves of the Amazon swords (*Echinodorus* spp.). For a subtle change, consider some of the *Hygrophila* species; their mass of small leaves creates a bush effect at the back of the aquarium. As well as leaf type, do not forget to make use of leaf colour. You can blend similar foliage tones into a background 'curtain' or use contrasting colours for impact.

The midground plants form the heart of the display and probably encompass the widest choice of suitable species. They should grow to about half the height of the aquarium, allowing you to view the background plants behind and create the tiered effect that looks so natural. Space out the midground plants a little to leave swimming space for the fish in the centre, yet provide sufficient cover for the most timid species. Positioning plants between items of hard landscaping conveys the impression that they are slowly invading the gaps. They will soon fill out,

softening the edges of rocks and bogwood and lending the display a more natural feel. Certain species of plants, including *Anubias* and *Microsorium*, do not grow in the substrate but on porous material, such as bogwood or suitable rock. These plants are ideal for the midground, as they combine hard decor with soft planting. The midground is also the ideal spot for a specimen plant, such as the Madagascan lace leaf plant (*Aponogeton madagascariensis*), that can become a centrepiece of the display.

The foreground should feature low-growing species that form a green carpet at the front of the display. *Glossostigma elatinoides* is an excellent example. Most of these species are rampant growers once established and will need regular pruning to keep them under control. It is safest to restrict yourself to one or two species to ensure that you do not end up with a tangle of different plants in the foreground. They will help to frame what is behind and achieve a natural look to the planted aquarium.

Floating plants add a finishing touch to complete the underwater scene. Many have trailing roots that provide welcome shade and refuge for timid fish within the aquarium. Floating plants can spread rapidly across the surface and need protection from bright lights that may scorch their leaves.

Background plants

Bacopa caroliniana
Giant Bacopa

When planted in groups, giant bacopa, from Central America, brings a delicate elegance to the back of the display, where the dense, small-leaved foliage easily hides any unsightly equipment. It can be propagated by cuttings planted around an existing group to increase its size. To thrive, the plant needs good lighting, a fine substrate and a regular supply of nutrients, as well as CO_2 fertilisation. In good conditions, it can grow to 20-40cm (8-16in).

Cabomba caroliniana
Green Cabomba

Probably the most common of the cabombas, this variety from Central and South America brings a lush, salad-green accent to the aquarium display. The fine bushy growth contrasts well with larger-leaved plants in the midground. Do not plant the stems in a bunch, but separate them out to allow light to reach the base of the plant. Most cabombas will produce sideshoots that can be nipped off with a pair of sharp scissors and replanted around the aquarium to make your cabomba display larger. *C. caroliniana* can grow up to 50cm (20in) and fills out the back of the aquarium.

Cabomba aquatica
Yellow Cabomba

The delicate leaves of this popular and commonly available plant from Central America add a feathery texture to the display. The leaf stalks are equally spaced along the stem, except near the tips, where the young growth is dense. Cabomba is often sold in bunches secured with a lead weight. Be sure to undo the weight and separate the stems before planting them 6cm (2.4in) apart, otherwise the lower leaves will not receive enough light and die off. This causes the lower stem to die back, break away and float around the aquarium. Cabomba requires regular feeding, strong lighting, and high temperatures to keep the growth healthy and compact. Reaches up to 40cm (16in).

Cabomba piauhyensis
Red Cabomba

Although not entirely red, this cabomba sports a wonderful mixture of flame colour shades from red to the lime-green of the new growth. It is usually the case that the most stunning plant needs the most care, and this is also true of Red Cabomba. To provide a stunning backdrop up to 40cm (16in) high, it needs excellent CO_2 fertilisation, very bright light and a regular supply of nutrients. Algae growth will quickly choke the delicate leaves, so make sure that it never gains a foothold. *C. piauhyensis* from Central and South America offers a challenge to the experienced aquarist; beginners should opt for the green variety, which is much easier to keep.

Crinum thaianum
Onion Plant

This stunning species has a very unusual appearance for an aquarium plant. The large, onion-shaped bulb sits half in the substrate, while ribbonlike leaves grow up towards the water surface. Once they get there, and this can even happen in a 1m (39in)-deep aquarium, they grow out along the surface. You will need to cut back some of these long leaves, as they may reduce light to other plants beneath them. However, it is important not to remove too many, or you will spoil the essential 'fountainlike' effect of the plant, which also provides valuable cover for the fish. Even though the onion plant is a background subject, choose a spot where some of the magnificent bulb can be seen.

Gymnocoronis spilanthoides
Spadeleaf Plant

This is another plant whose arrangement of relatively small leaves can create a wonderful backdrop for the display. Pairs of leaves grow from a central stem at 90° to the pair below to maximise the light absorbed by the plant. The spade-shaped green leaves can form very quickly; in ideal conditions, *G. spilanthoides* can reach 50-60cm (20-24in) and needs regular pruning to prevent it becoming too rampant. If bought in bunches, separate each of the stems before planting so that light can reach the lower leaves.

Hygrophila polysperma 'Big Leaf' grows larger than the standard form of this hardy plant.

Hygrophila polysperma
Dwarf Hygro

Dwarf hygrophila, from India, is the most commonly available *Hygrophila* species and ideal for the beginner, as its rapid growth and dense foliage reassure the newcomer that things can develop quickly in the aquarium. Regular pruning helps to generate new growth and keeps the plant healthy. As it is such a popular plant, cultivars such as 'Rosenervig' have been introduced. The white veins in its leaves add a structural dimension to the plant's appearance. The leaves also turn an attractive pink shade as they reach bright light toward the surface.

Hygrophila difformis
Water Wisteria

This broad plant from India, Thailand and Malaysia has delicate, finely branched leaves that billow elegantly in the flow of water from a filter. To give of its best it needs regular feeding, good lighting and sufficient space. If it is too crowded, the lower part of the plant cannot receive enough light and quickly disintegrates. It will grow to a height of 50cm (20in) and is ideal for masking unsightly aquarium equipment.

Hygrophila guianensis
Hygro

The large leaves of *H. guianensis* add an unusual shape to the display. Try several stems growing together to make a wall of foliage, 25cm (10in) high. However, it is the most difficult species to keep in an aquarium environment, as it needs plenty of nutrients at the roots and strong lighting for its leaves to do well. Beginners will find the *polysperma* species much easier to keep.

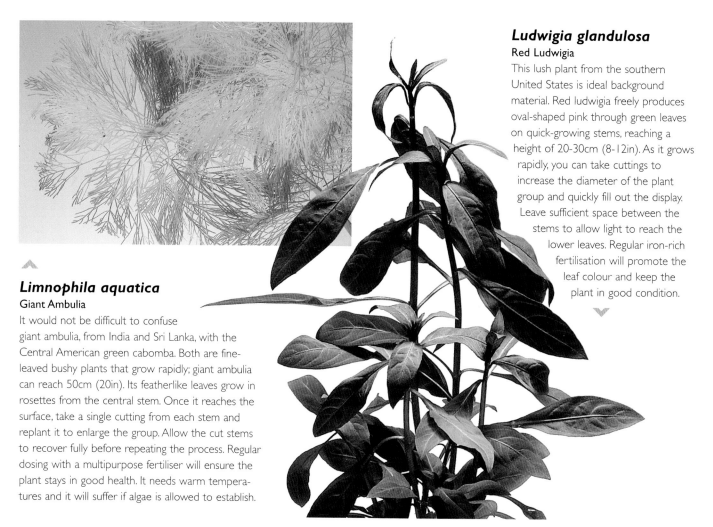

Ludwigia glandulosa
Red Ludwigia

This lush plant from the southern United States is ideal background material. Red ludwigia freely produces oval-shaped pink through green leaves on quick-growing stems, reaching a height of 20-30cm (8-12in). As it grows rapidly, you can take cuttings to increase the diameter of the plant group and quickly fill out the display. Leave sufficient space between the stems to allow light to reach the lower leaves. Regular iron-rich fertilisation will promote the leaf colour and keep the plant in good condition.

Limnophila aquatica
Giant Ambulia

It would not be difficult to confuse giant ambulia, from India and Sri Lanka, with the Central American green cabomba. Both are fine-leaved bushy plants that grow rapidly; giant ambulia can reach 50cm (20in). Its featherlike leaves grow in rosettes from the central stem. Once it reaches the surface, take a single cutting from each stem and replant it to enlarge the group. Allow the cut stems to recover fully before repeating the process. Regular dosing with a multipurpose fertiliser will ensure the plant stays in good health. It needs warm temperatures and it will suffer if algae is allowed to establish.

Myriophyllum aquaticum
Brazilian Milfoil

Milfoil could be described as the feather boa of the aquarium plant portfolio. It can add a wonderful fluid texture to the background of the display, as its many fine leaves often completely hide the stems. However, this fine foliage can also be the plant's downfall, as it will trap any passing debris. Efficient mechanical filtration is essential. Keep the plant healthy with regular feeds and strong lighting. Resist the temptation to take stem cuttings; wait until the plant produces a sideshoot and prune this to expand the area of your plant. This South American species can grow to 50cm (20in).

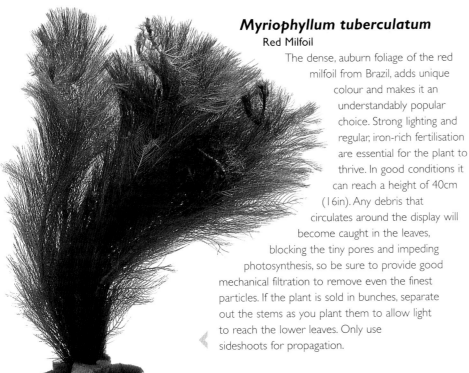

Myriophyllum tuberculatum
Red Milfoil

The dense, auburn foliage of the red milfoil from Brazil, adds unique colour and makes it an understandably popular choice. Strong lighting and regular, iron-rich fertilisation are essential for the plant to thrive. In good conditions it can reach a height of 40cm (16in). Any debris that circulates around the display will become caught in the leaves, blocking the tiny pores and impeding photosynthesis, so be sure to provide good mechanical filtration to remove even the finest particles. If the plant is sold in bunches, separate out the stems as you plant them to allow light to reach the lower leaves. Only use sideshoots for propagation.

Rotala macrandra
Giant Red Rotala

Ruffles of leaves ranging in colour from pink to deep red, with a hint of green on the upper sides, make this Indian plant a superb addition to the aquarium. However, as you might expect from a plant with mostly red foliage, only strong lighting and iron-rich fertilisation will keep the colours bold on new and existing growth. Furthermore, it is delicate and should be handled very gently, as both the leaves and stems can be easily damaged. The lower leaves can die off if they have insufficient light, so leave space between the stems when planting. In good conditions, giant red rotala can grow to 50cm (20in).

Nesaea crassicaulis

This very striking plant for the rear of the aquarium has pairs of lush green leaves growing on strong stems to a height of 40-50cm (16-20in). The tips of the new growth are often pinkish; good lighting and regular, iron-rich fertilisation will enhance this attractive coloration. The plant holds its elongated leaves horizontally from the stem to gather the most light, and this tendency gives the plant a very strong structural appearance. When planting, leave sufficient space between the stems to allow light to reach the base. N. crassicaulis prefers slightly soft water. If you are not certain you can provide the conditions this West African plant requires, then Hygrophila polysperma might be a safer option.

Sagittaria subulata
Needle Sagittaria

An ideal plant for the beginner, the needle sagittaria from the eastern United States will grow in most aquarium conditions, reaching a height of 30cm (12in). Its long, thin leaves grow in dense bundles to fill out the background. It will thrive in good light, although in deeper-than-average aquariums it may do better in the midground. Needle sagittaria produces runners that spread out to form a curtain of plants that can be used as a 'divider' between other plants if your design requires it.

Shinnersia rivularis
Mexican Oakleaf

As the common name implies, this plant's leaves resemble those of the terrestrial oak tree. They grow on tall stems that become leggy in poor light, so provide very good illumination to maintain a bushy shape. Regularly pruning the top growth prevents the development of large leaves that will cut out light to other plants below. Mexican oakleaf can grow up to 60cm (24in); to increase the display, prune and replant the side growths.

Vallisneria asiatica var. biwaensis
Corkscrew Val

Growing to a height of 35cm (14in), the tightly twisted leaves of this striking plant from Japan make a stunning backdrop in the aquarium. Corkscrew val spreads rapidly by means of daughter plants produced on runners and will need controlling to ensure that it does not become too invasive. Do not confuse this species with its smaller cousin, *Vallisneria tortifolia*, whose leaves are not as twisted and which does not grow as tall.

Vallisneria gigantea
Giant Val

If a bold backdrop is what you are after, then the giant val from New Guinea is the plant you need. The vigorous ribbonlike leaves, up to 1m (39in) tall, soon reach the surface and spread horizontally. Careful management will prevent the plant from becoming dominant. As the plant is so vigorous, it requires strong lighting and good-quality, iron-rich fertilisation. Yellowing of the leaves is a sure sign that there is an insufficient supply of light and iron. Giant val spreads via daughter plants on runners and thrives in cool, moderately hard water.

Vallisneria spiralis
Straight Val

At 60cm (24in), straight val is shorter than the giant species and suitable for most aquarium displays. Once the thin, ribbonlike leaves reach the surface, they grow outwards to make the best use of the available light. Propagation is via runners from the main plant, and these may need controlling to prevent the plant becoming invasive. In common with most Vallisneria species, established plants send up twisted flower spikes that open just above the water surface. Good lighting and iron-rich fertilisation are essential for this plant to thrive.

Vallisneria spiralis 'Tiger' has thinner, even more elegant leaves than the natural form.

Midground plants

Ammannia gracilis
Delicate Ammannia

Ammannia is a very graceful plant from Africa, with thin ribbonlike leaves on a cylindrical stem. Both the leaves and the stem are the same reddish colour. As the upper growth can cut out light to the base of the plant, always grow it behind some bushy foreground plants to hide the lower parts. When planted in a group, ammannia can create a very pleasing forest effect. Under ideal conditions it can grow to 50cm (20in), but 25-30cm (10-12in) is more realistic. Provide plenty of iron-rich substrate and bright lighting.

Alternanthera reineckii

A plant of colourful contrasts is what you will find with alternanthera. The upper surface of the leaves is olive green, while the underside is a striking pink. The leaf pairs grow from a central red stem. Planting A. reineckii in bushy groups creates a more natural effect than planting single stems. An iron-rich substrate will help to enhance the red coloration of this South American specimen, while bright lighting will encourage the plant to grow strongly. Its maximum height in optimum conditions is 50cm (20in).

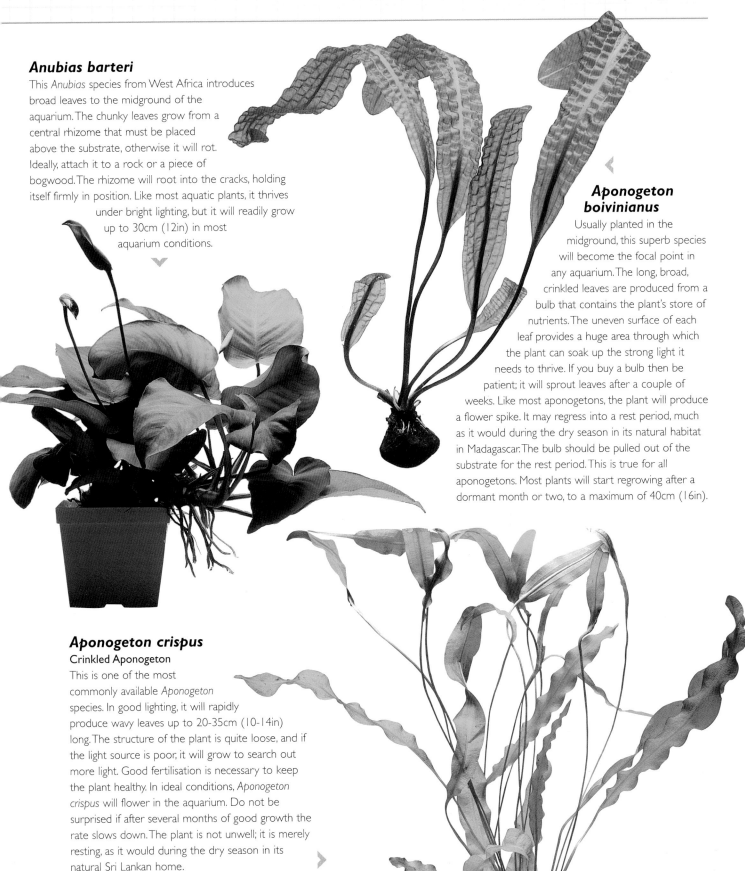

Anubias barteri

This *Anubias* species from West Africa introduces broad leaves to the midground of the aquarium. The chunky leaves grow from a central rhizome that must be placed above the substrate, otherwise it will rot. Ideally, attach it to a rock or a piece of bogwood. The rhizome will root into the cracks, holding itself firmly in position. Like most aquatic plants, it thrives under bright lighting, but it will readily grow up to 30cm (12in) in most aquarium conditions.

Aponogeton boivinianus

Usually planted in the midground, this superb species will become the focal point in any aquarium. The long, broad, crinkled leaves are produced from a bulb that contains the plant's store of nutrients. The uneven surface of each leaf provides a huge area through which the plant can soak up the strong light it needs to thrive. If you buy a bulb then be patient; it will sprout leaves after a couple of weeks. Like most aponogetons, the plant will produce a flower spike. It may regress into a rest period, much as it would during the dry season in its natural habitat in Madagascar. The bulb should be pulled out of the substrate for the rest period. This is true for all aponogetons. Most plants will start regrowing after a dormant month or two, to a maximum of 40cm (16in).

Aponogeton crispus
Crinkled Aponogeton

This is one of the most commonly available *Aponogeton* species. In good lighting, it will rapidly produce wavy leaves up to 20-35cm (10-14in) long. The structure of the plant is quite loose, and if the light source is poor, it will grow to search out more light. Good fertilisation is necessary to keep the plant healthy. In ideal conditions, *Aponogeton crispus* will flower in the aquarium. Do not be surprised if after several months of good growth the rate slows down. The plant is not unwell; it is merely resting, as it would during the dry season in its natural Sri Lankan home.

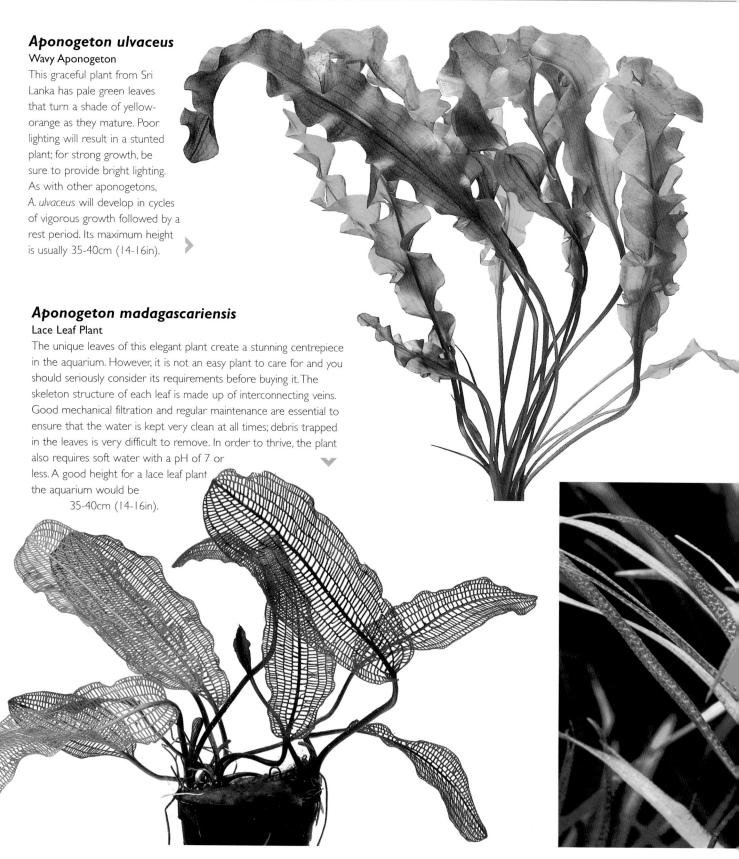

Aponogeton ulvaceus
Wavy Aponogeton

This graceful plant from Sri Lanka has pale green leaves that turn a shade of yellow-orange as they mature. Poor lighting will result in a stunted plant; for strong growth, be sure to provide bright lighting. As with other aponogetons, *A. ulvaceus* will develop in cycles of vigorous growth followed by a rest period. Its maximum height is usually 35-40cm (14-16in).

Aponogeton madagascariensis
Lace Leaf Plant

The unique leaves of this elegant plant create a stunning centrepiece in the aquarium. However, it is not an easy plant to care for and you should seriously consider its requirements before buying it. The skeleton structure of each leaf is made up of interconnecting veins. Good mechanical filtration and regular maintenance are essential to ensure that the water is kept very clean at all times; debris trapped in the leaves is very difficult to remove. In order to thrive, the plant also requires soft water with a pH of 7 or less. A good height for a lace leaf plant the aquarium would be 35-40cm (14-16in).

Barclaya longifolia
Orchid Lily

Often bought as a dour-looking corm, *Barclaya longifolia* will reward you with a wondrous display of brightly coloured young leaves that become darker green on top and red-brown beneath. The slender arching leaf stalks add a different shape to the display. As the plant grows from a corm, it needs to replace the nutrients it uses up, so provide good fertilisation, especially CO_2. If you can also supply good-quality bright lighting, *Barclaya longifolia*, from East Africa, is well worth considering as a show plant for the midground. It will grow to 35cm (14in).

Blyxa japonica
Japanese Rush

This Asian plant is difficult to care for, but in ideal conditions it can form a dense mass of leaves. To give it a prolonged, successful life, provide soft water with a low pH, bright lighting and high levels of CO_2 fertilisation. Although it only grows to a maximum height of 8cm (3.2in), the ribbonlike leaves will create a bush effect in the centre of the display.

Cryptocoryne balansae

C. balansae from Thailand is a very different-looking member of the *Cryptocoryne* group. The ribbonlike leaves with a crinkled, indented appearance contrast well with other plants and add a different texture to the display. The plant grows well in most conditions, but bright lighting will bring out the best in it. The leaves can grow 40cm (16in) tall in ideal conditions, but 10cm (4in) is the more likely aquarium height.

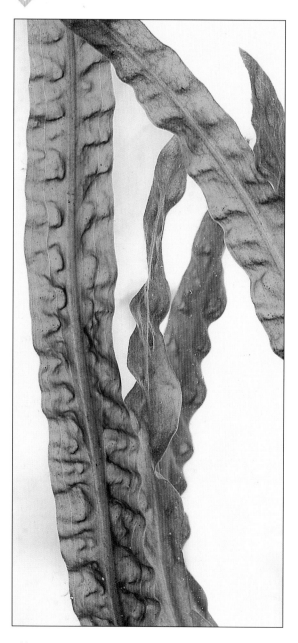

Cryptocoryne pontederiifolia

This is one of the hardiest *Cryptocoryne* species and an easy plant to grow. Given bright lighting and CO_2 fertilisation, it will grow up to 25cm (10in) high with a similar width. The bold leaf shape of this plant from Sumatra and Borneo will create a central focal point in the aquarium.

Cryptocoryne cordata
Giant Cryptocoryne

The large oval leaves of this distinctive cryptocoryne from Thailand are an unusual colour for an aquarium plant, being reddish-brown on top and dark red beneath. In the right conditions, the leaves can grow up to 15cm (6in) long and 6cm (2.4in) wide, hence the description 'giant' in the plant's common name. To generate these large leaves, the plant requires a good nutrient supply from both CO_2 and a nutrient-rich substrate.

Cryptocoryne undulata
Undulate Cryptocoryne

Dark green undulating leaves held on contrasting red-brown leaf stalks are features of this popular midground plant from India. The leaves can reach 30-35cm (12-14in) long, but poor light levels will cause it to produce smaller, pale green leaves. Good lighting plus a nutrient-rich substrate will encourage strong growth.

Echinodorus cordifolius
Radicans Swordplant

The almost circular leaves of this distinctive plant from North America and Mexico add yet another dimension to a planted display. Single leaves are borne on a succession of leaf stalks up to 40cm (16in) tall. Removing older leaves will keep the plant tidy and encourage new growth. Again, this swordplant requires a good supply of iron-rich fertiliser for healthy growth. The plant produces a flower spike that unfurls to reveal small white flowers above the water level.

Echinodorus bleheri
Broadleaved Amazon Swordplant

This easy-care plant from South America is probably the most common Amazon Swordplant found in aquarium shops. When planting, leave plenty of space around it for future growth, and place undemanding foreground plants in front to hide the base, which can look messy with age. The vigorous growth needs to be fuelled by bright lighting and a constant supply of iron-rich fertilisers, plus CO_2. In ideal conditions, this plant will reach 50cm (20in) high and produce flower spikes and runners that enable it to spread around the aquarium.

Large leaves with strong veins make a bold statement in the aquarium display.

Echinodorus macrophyllus
Large-leaved Amazon Swordplant

Since this statuesque Amazon swordplant from Guyana and Brazil can reach 50cm (20in) tall and produce leaves 20cm (8in) long, it is really only suitable for the most spacious aquariums. The large oval leaves will block out light to other plants, although regular pruning should keep the plant in check. As there are so many swordplants to choose from, however, a smaller species would be a better option for medium-sized aquariums.

Echinodorus martii
Ruffled Amazon Swordplant

The elongated leaves of *E. major* are ruffled around the edges. To thrive, the plant requires iron-rich fertilisation and bright lighting. This Brazilian plant, which can grow to 50cm (20in), is ideally suited to the larger aquarium, where it makes an excellent centrepiece.

Echinodorus osiris
Red Amazon Swordplant

In *E. osiris*, one of the most stunning Amazon swordplants, the young red leaves revert to the more traditional green as they mature. Another notable feature of this species from Brazil is the clearly visible structure of veins in the leaves. In common with the other swordplants, it requires bright lighting and iron-rich fertilisation to reach its maximum height of 40-50cm (16-20in).

Echinodorus 'Red Flame'

This is another cultivated variety that is popular with aquarium plant growers. The stunning green-and-red mottled leaves produce a striking display and add variety to any aquarium. The red speckled effect is evident throughout the life of each leaf. The plant, originally from South America, will grow up to 40cm (16in) tall, making it an ideal focal point in the aquarium.

Echinodorus palaefolius var. latifolius

The oval green leaves of this specimen swordplant from Brazil are borne on stalks that can be 30-35cm (12-14in) tall. Its vigorous growth rate, however, ensures that new leaves are constantly emerging to cover up the base of the stalks. Give this plant plenty of room and avoid placing light-dependent species under its generous canopy. Bright light and iron-rich fertilisers will maximise its potential, although it will grow in most conditions.

Echinodorus parviflorus

As one of the more compact species of Amazon swordplant, E. parviflorus from South America is ideal for most aquariums. The short leaf stalks carry compact leaves that are elongated to a pointed tip. Used in the midground, it will hide the unsightly stalks of taller plants at the back of the aquarium. Provide the usual high-iron fertilisation and bright lighting to ensure that the plant reaches its potential mature height of 25cm (10in).

Echinodorus 'Rubin'

The leaves of E. 'Rubin', one of the most stunning cultivated hybrids available for aquarium keepers, vary in colour from red to green as the plant grows. The leaf veins are very visible, making this cultivar an excellent focal point in the centre of the aquarium. The mature leaves have a tendency to spread as opposed to growing up towards the light, so give the plant plenty of room. Bright lighting and iron-rich fertilisers will keep it in good health. It can grow to 50cm (20in).

Boldly marked leaves in shades of red and green make this plant a stunner.

Echinodorus uruguayensis
Uruguay Amazon Swordplant

With numerous thin leaves measuring up to 30cm (12in) long, this swordplant from Southern Brazil and Uruguay is an interesting departure from the usual broadleaved species. As it grows, it creates a fan effect that looks excellent in the midground. When young, the leaves are often reddish in colour, making a wonderful contrast with the adult green leaves. Bright lighting and an iron-rich, fine substrate will keep the plant in good condition. It can tolerate slightly cooler water conditions than other swordplants.

Echinodorus 'Red Special'

The name of this cultivated swordplant variety clearly explains its appeal. The copper-red foliage is markedly different to the usual green swordplant species and makes a great addition to any aquarium. The large, oval leaves often spread out, so give them plenty of room when planting. Bright lighting, iron-rich fertiliser and the addition of CO_2 all help to ensure that the plant looks its best and remains the 40cm (16in)-tall showpiece of your aquarium display.

Heteranthera zosterifolia
Stargrass

Borne on thin stems, the leaves of *H. zosterifolia* from Brazil are produced in a starlike arrangement. The plant's main requirement is good lighting and given this it will remain compact, forming dense bushy growth up to 40-50cm (16-20in) high in the middle of the aquarium.

A lack of light will cause it to grow to the surface in search of more illumination. Once it has found a level with sufficient light, it will throw out more sideshoots and become bushier. Stargrass provides an excellent contrast to larger-leaved plants.

Above: *The branching stems of* Hydrocotyle leucocephala *add a vertical accent to this display.*

Hydrocotyle leucocephala
Brazilian Pennywort

With circular leaves measuring 3-5cm (1.2-2in) across, pennywort makes a distinctive addition to the aquarium. It is a fast grower, up to 50-60cm (20-24in) tall, and needs regular pruning to keep it under control and prevent it smothering other plants. Bright light is the most important requirement. The shoots double as runners; if they break off, they will readily form new plants.

Ludwigia arcuata

The dense growth of this *Ludwigia* species from North America makes it ideal for the midground of the aquarium. The leaves are narrower than those of *L. repens* and the plant makes a pleasing contrast when placed next to its cousin. The leaves require strong lighting and iron-rich fertilisation to develop a reddish hue. In low light, the leaves will stay a pale green and the plant will not thrive. To maintain the dense appearance, trim off any leggy growth.

It will grow to 25-50cm (10-20in) in height.

Microsorium pteropus 'Tropica'

In this cultivated variety of Java fern, the deeply cut leaf edges appear to form lobes. The plant's requirements are the same as for normal Java fern.

Ludwigia repens
Narrow-leaved Ludwigia

In its different forms, this small-leaved plant from Central and North America brings a variety of leaf shapes and colour to the aquarium. In *L. repens*, the leaves are dark green to brown on top, with a reddish-green underside. Being a fast-growing plant, it will quickly make an impact in the aquarium. As it grows, the plant produces pairs of leaves along the stem. At each junction between a leaf and the stem a bud develops that can grow to form an entirely new shoot with its own set of leaves. The gaps between the leaves are governed by the amount of light the plant receives; the greater the light intensity, the closer the leaf pairs are together. It can reach a height of 50cm (20in) and may need controlling to stay in scale with other midground plants.

Microsorium pteropus
Java Fern

This is one of the easiest plants to keep in the aquarium.
In its natural environment in Southeast Asia, it grows both
submerged and above water on rocks and
wood at the edge of streams and rivers.
If sold unattached, be sure to tie it to a rock or
a piece of bogwood, using cotton or thin fishing line.
Older leaves often have plantlets growing on them
and these can be used for propagation. In nature, they
form new plants that fall from the parent plant as the old
leaves die off. M. pteropus will grow slowly to 25cm (10in) and
form a bushy plant. It will thrive in a wide range of aquarium
conditions and does not need bright light.

Microsorium pteropus 'Windeløv'

An even more finely cut edge gives the tip of each leaf
of this variety a feathery appearance. In addition, the
leaves of this highly sought-after aquarium plant are
generally more compact than those of the normal Java
            fern, making it suitable for the smaller aquarium.

Nymphaea stellata
Red and Blue Water Lily

Being a smaller-leaved species than N. lotus, this water lily from
India is a better choice for the more compact aquarium. The
leaves are not mottled, but range from brownish pink to red.
Removing long-stemmed leaves helps to keep the plant compact.
Plant the bulb with the crown showing; if buried too deeply,
it has a tendency to rot and die off.

Above: *The leaves of* Nymphaea lotus *'Zenkeri' vary in colour and pattern between specimens.*

Nymphaea lotus
Tiger Lotus

Water lilies are popular the world over and this East African species is ideal for the aquarium keeper. All water lilies like to grow to the surface to gain maximum benefit from the available light, but this is not always desirable in the aquarium.
Removing any leaves that look as though they will reach the surface – and this can happen in as little as three to four days – will encourage the plant to produce more submerged growth. The mottled green-and-red leaves can reach 18-20cm (7-8in) across, producing a bold aquarium display. Give N. lotus plenty of room in the middle of the aquarium so that it does not cut out light to other plants. All the leaves grow from a tuber, which should be planted in the substrate with just the top of the crown showing.

Sagittaria platyphylla
Giant Sagittaria

Although *S. platyphylla* can be grown as a foreground plant, it often reaches 20cm (8in) tall, so is better suited to the midground. The word 'Giant' in the common name refers to the broad leaves that grow in a fan shape and make a bold contrast with other plants across the middle of the display. To thrive and spread, *S. platyphylla* requires regular iron-rich fertilisation as well as bright, good-quality lighting.

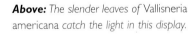

Above: *The slender leaves of* Vallisneria americana *catch the light in this display.*

Vallisneria americana
Jungle Val

V. americana is ideal for planting at the sides of the midground area. Here, the ribbonlike leaves that twist as they grow to 50-100cm (20-40in) will effectively hide any aquarium equipment. This North American plant spreads by producing runners and the daughter plants will form a dense patch under ideal conditions. Given bright light, it will also produce flower shoots that wind their way to the surface and bear a small white flower.

Rotala rotundifolia
Dwarf Rotala

In contrast with most midground subjects, this small-leaved plant from Southeast Asia makes an ideal bushy group when several specimens are planted together. The leaves and stem are a lush green colour; in good light, the leaves may acquire a pinkish hue towards the tips. In addition to good-quality bright lighting, a regular supply of nutrients will help it thrive.

Foreground plants

Anubias barteri var. *nana*
Dwarf Anubias

This is probably the most popular *Anubias* species.
It is also one of the smallest and lends itself beautifully
to planting at the front of the display. The *nana*
variety, from West Africa, has waxy round
leaves on short stems that grow from the
creeping rhizome. The plant's stocky appearance
seems to deter the unwanted attentions of herbivorous
fish. This undemanding species grows slowly to a height of
12cm (4.7in) and can be bought growing over a rock or a
piece of bogwood, which adds extra interest to the display.

Anubias angustifolia 'Afzelii'
Narrow-leaved Anubias

Anubias species are among the most robust and undemanding plants you
can add to your display. They grow from a rhizome and produce plenty of
leaves. 'Afzelii', from West Africa, has crisp, green, pointed leaves that form
a tough, bushy plant generally ignored by fishes. It is best positioned at the
side of the foreground area as it can reach 20cm (8in) tall, although this
could take some time. If it becomes a little big for the foreground, move it
back slightly to blend with the midground plants. Like other *Anubias*
species, it can be attached to rocks
or bogwood. This excellent
starter plant for the
aquarium will thrive
with CO_2
fertilisation.

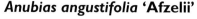

Bacopa monnieri
Dwarf Bacopa

With regular pruning and in good-quality, bright light, dwarf bacopa will develop into a compact, bushy foreground plant less than 10cm (4in) high. It produces small clusters of thick, oval leaves. However, in low-light aquarium conditions, growth will be long and leggy, with larger spaces between the leaves.

Cryptocoryne albida

The slender leaves of this beautiful cryptocoryne can vary in colour from green to reddish brown, and contrast well with other foreground plants. To create a display with visual impact, plant it in a group of at least three. C. albida, from Thailand, requires regular fertilisation and bright, good-quality light if it is to remain compact. In poor light, growth will be leggy. Like all cryptocorynes, it appreciates a substrate augmented with laterite and in the long term will benefit from fertiliser tablets close to its roots. It will reach 25-30cm (10-12in) in height.

Cladophora aegagropila
Moss Balls

These slow-growing spherical balls of algae add a hugely original and interesting plant form to the foreground. Buoyed up by gases released during photosynthesis and respiration, they float to the surface, then sink again when the gases are released.

Cryptocoryne willisii

This tiny Sri Lankan cryptocoryne, which grows from
5-12cm (2-5in), is another ideal foreground subject. It has
broader leaves than *C. parva* and also spreads to
fill a space across the front of an aquarium. Like
all small plants, it requires strong lighting but is
otherwise easy to care for. Provide regular
fertilisation to ensure that it remains healthy.

Cryptocoryne beckettii
Beckett's Cryptocoryne

Although it only grows to a maximum of 15cm
(6in), this attractive cryptocoryne from Sri Lanka
produces large, deep green leaves that complement
the reddish-brown stalks. This excellent foreground
species looks best when
planted in a small group and
will bush out in time. Make sure
that taller plants behind do not
block out the light, otherwise the plants
will suffer. Provide a good-quality
substrate and add nutrients in the form of
root tablet fertilisers. Corydoras and other
small fish species enjoy the cover
created by the broad leaves.

Cryptocoryne parva
Dwarf Cryptocoryne

Once established, this beautiful
delicate, thin-leaved cryptocoryne
from Sri Lanka creates a 'lawn'
across the entire front of
the aquarium.
However, to
achieve this effect you will need
to provide a nutrient-rich
substrate and good-
quality, bright lighting.
C. parva grows to a maximum
height of 5cm (2in) and makes
an excellent complēment to
its larger relatives.

Didiplis diandra

This North American species likes to grow towards the surface to a height of 25-35cm (10-14in) and therefore may seem an unusual inclusion as a foreground plant. However, it can be kept smaller with regular pruning to make it produce bushier sideshoots and a hedgelike appearance. Its delicate, feathery leaves help it stand out from most other aquatic plants. In good lighting, the uppermost leaves change colour from green to rusty brown. It requires bright light, soft water and regular feeding with iron-rich food. If any long tips of growth break off, simply push them into the substrate.

Cryptocoryne walkeri

Being a slightly taller cryptocoryne, *C. walkeri* makes an excellent side foreground plant to soften the corners of the aquarium. It is one of the most commonly available species and its dark green leaves with brown undersides are popular with aquarium keepers. When planted together with a carpet of *C. parva*, it can create a complete foreground. *C. walkeri*, from Sri Lanka, grows up to 12cm (4.7in) high and looks best planted in a small group. It is easy to grow and will benefit from regular fertilisation.

Echinodorus tenellus
Pygmy Chain Swordplant

Originating from North and South America, this is one of the most popular foreground plants but requires good conditions if it is to thrive. In bright lighting and given a fine substrate, it will spread to form a carpet only 8cm (3.2in) high across the front of the display. Any lobe-shaped leaves on a newly acquired plant are an indication that it has been grown above water. However, these leaves will die back once the plant is submerged, to be replaced by thinner, aquatic foliage. If the plant becomes too invasive, simply cut off the runners and lift out the unwanted pieces.

Eleocharis acicularis
Hairgrass

This grasslike aquatic plant is readily available from most aquatic stores. In dim conditions it will grow leggy, but under bright light and with regular doses of fertiliser, it bushes outwards, creating an interesting texture in the display. Be sure to include some corydoras catfish in the aquarium to eat or dislodge any debris that becomes trapped in its fine leaves. If these become clogged, the plant may die back. Can grow to 25cm (10in), but usually 15-20cm (6-8in).

Fontinalis antipyretica
Willow Moss

This interesting small plant from North America, Europe, Asia and North Africa makes a distinctive addition to the display. It produces numerous, small, sail-like leaves 0.5cm (0.2in) long on thin stems and attaches itself by its roots to rocks and bogwood. Happier in cooler conditions, it enjoys moving water from the filter outflow. When you first buy willow moss, handle it gently and anchor it carefully to a piece of wood or rock in the foreground using fine fishing line. In bright light, it will spread to create a mat that covers the line and rapidly looks as though it has always grown there.

Eusteralis stellata
Star rotala

Although this stunning plant from Australia and Asia can grow to 40cm (16in) tall, regular pruning and the correct conditions can keep it squat and bushy, making it an excellent foreground subject. To keep the starlike rings of foliage looking their best, provide strong lighting, high levels of dissolved carbon dioxide and a substrate rich in nutrients. Although the star rotala makes high demands, it rewards in equal measures.

Glossostigma elatinoides

Originating from Australia and New Zealand, Glossostigma must be the ultimate foreground carpet plant. Providing it receives strong lighting and CO_2 fertilisation, it will form a dense mat across the entire foreground area, softening any hard edges and corners. The effect is of a display that has been cut out of a river bed. It rarely grows more than 1.5cm (0.6in) high. If the plant is growing healthily, you can often observe how the small, lobed leaves trap escaping oxygen produced during photosynthesis. Arrange the plants in small groups to start with and watch as they blend into one mass as they become established.

Hemianthus micranthemoides
Pearlweed

At 15-20cm (6-8in), *H. micranthemoides* from Cuba and southeastern United States is slightly taller than dwarf helzine, and can be kept bushy with regular trimming. Try to keep it free of debris, as the fine, lobed leaves easily become clogged with suspended particles. To create a carpet effect you will need several individual plants, but these will soon blend together to give the impression of a single, larger plant. It is vital to provide bright, good-quality lighting.

Hydrocotyle sibthorpioides
Dwarf Pennywort

Borne on low-growing runners, the round leaves of this *Hydrocotyle* species introduce a different foliage shape to the foreground. This Southeast Asian plant requires bright light if it is to do well, but in the right conditions will grow reasonably quickly to cover the substrate. For maximum effect, plant a group of three or more. It will reach 12cm (4.7in) high.

Hemianthus callitrichoides
Dwarf Helzine

This compact Central American plant creates a dense mat of small round leaves 3-5cm (1.2-2in) high. It has neither the growth rate nor the marauding tendencies of *Glossostigma*. Dwarf helzine can form an excellent contrast to other foreground plants, such as the smaller cryptocorynes. All *Hemianthus* species require a good supply of nutrients and bright lighting, without which they soon suffer and die off.

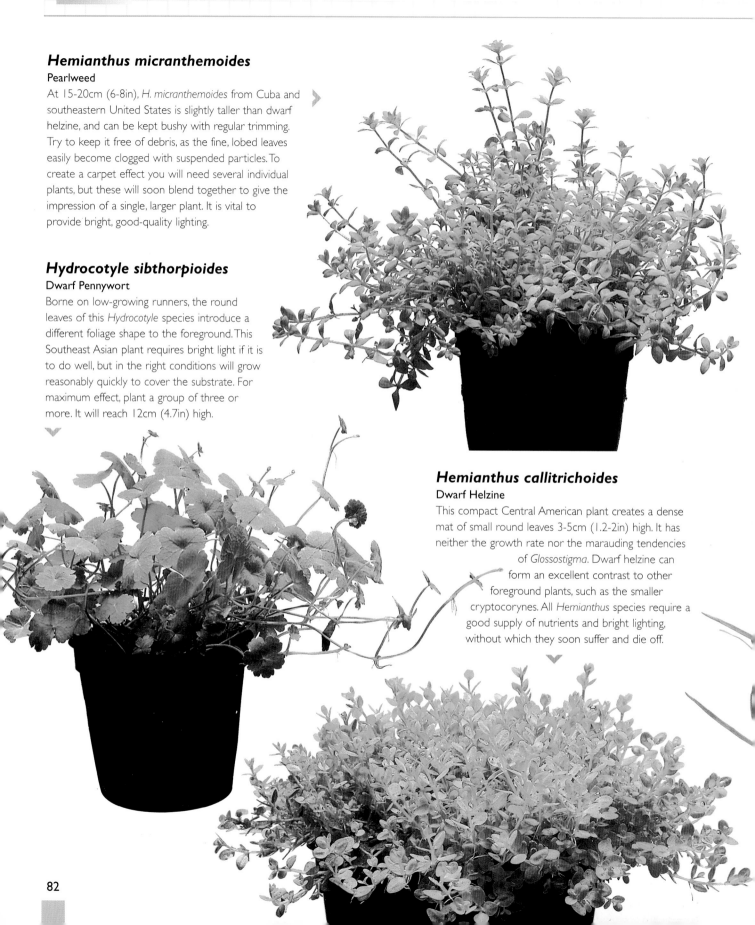

Micranthemum umbrosum
Helzine

Unlike *Glossostigma*, which grows in an organised manner, *M. umbrosum* provides a jumbled display of foliage in the foreground. It is an attractive, squat plant from Central America that produces copious amounts of very small, round leaves on cresslike stems up to 30cm (12in) tall. It thrives in bright light, where it remains more compact. It can grow as a floating plant, but may need regular cropping to prevent it shading the aquarium below.

Lilaeopsis mauritiana
Grassplant

This tufty little plant from the Indian Ocean island of Mauritius spreads across the foreground, complementing the broader-leaved plants behind and providing cover for smaller fish. Corydoras catfish provide a useful service by keeping it clear of debris. Strong lighting and a fine substrate will ensure healthy growth. The plant grows 5-10cm (2-4in) tall, but regular trimming will keep it under control. Cutting it back to different heights will result in a more natural look.

Monosolenium tenerum
Pellia

Originating from Asia, this stunning, small-leaved, fernlike plant is sold attached to rocks or purpose-made 'plant stones'. In good lighting and with reasonable CO_2 fertilisation, this 'living fossil' soon creates a pad of foliage in which fish love to rummage. *Monosolenium* does not have true leaves, but grows thalli, which look like webbed pads that fork as they grow. Small pieces break off, lodge elsewhere in the aquarium and start new plants. This unusual species can be tiered towards the back of the aquarium to create an impressive display.

Floating plants

Pistia stratiotes
Water Lettuce

This excellent floating plant is best suited to the large aquarium. It produces velvety, lettucelike leaves above the surface and long, feathery roots below. The plant absorbs nutrients through the roots, which also provide welcome shade and sanctuary for fish. Good ventilation above the aquarium is required if the plants are to do well, but otherwise it is easy to care for.

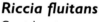

Limnobium laevigatum
Amazon Frogbit

Resembling mini lily pads, these chunky, small, round-leaved plants from Brazil float around the aquarium in the flow of water. They can easily spread to cover a large area of the tank, so be prepared to keep them under control. Daughter plants produced on runners soon provide a clump of leaves under which surface-dwelling fish can hide.

Riccia fluitans
Crystalwort

Unusual, but hardy and undemanding, this simple floating plant spreads by division of the growing tips. Careful management may be required if it becomes too vigorous, as it will cut out light to the plants beneath it. It is a very adaptable plant that will do well in most aquariums.

Salvinia natans
Salvinia

The leaves of this *Salvinia* species from Europe, Asia and North Africa have a crinkled appearance, which helps to distinguish it from *S. auriculata*. Like all floating plants, it provides excellent cover for surface-swimming fish. The feathery roots extract nutrients from the water; good fertilisation will help the plant stay healthy.

In *Salvinia oblongifolia* (inset) the leaves are more oval and elongated, but the plant's requirements are similar to those of other *Salvinia* species.

Salvinia auriculata
Salvinia

This fast-growing plant from Central and South America produces pairs of velvety leaves above the water surface as it drifts around the aquarium. It grows by absorbing dissolved nutrients from the water through its roots. In time, it can create a dense carpet across the surface, so make sure it does not get out of control and block out all the light to the plants below. Without sufficient ventilation above the tank, bright halide lighting may scorch the leaves.

LIGHTING THE AQUARIUM

All life on our planet ultimately relies on light energy from the sun to survive. Plants use sunlight for photosynthesis to build simple sugars as the starting point of all food chains. In the 'artificial' environment of the aquarium, the lighting system must perform the roles of providing light energy for plants to flourish and to illuminate the display so that fish can feel at home and we can see and enjoy the combined result.

What is light?

Visible light occupies a relatively narrow band in a wide span of electromagnetic energy that stretches from low-energy radio waves at one end to high-energy gamma waves at the other. The only difference between these forms of energy is their wavelength, commonly measured in nanometres (nm) – billionths of a metre. The visible light spectrum starts at infrared (700nm) and extends to ultraviolet (400nm). Between these are the rainbow colours we are familiar with, ranging from red and orange through yellow to green and blue. Combined, they make white light.

This explanation might seem rather scientific, but it important to understand how light 'works' so we can choose the most efficient lighting system for the aquarium display.

What light do plants need?

Most plants use the green pigment chlorophyll to trap sunlight for photosynthesis. Chlorophyll absorbs mainly red and blue light. (It makes plants look green simply because it reflects and does not absorb green light.) In the red part of the spectrum, the wavelengths between 650 and 680nm are the most efficiently trapped by chlorophyll.

Land plants receive 'equal shares' of all the wavelengths of light from the sun. However, submerged aquatic plants must cope with the light filtering effects of water. The longer wavelengths at the red

Recreating natural light in the aquarium

Left: *In some parts of the Amazon River Basin the water is crystal clear and the strong tropical sunlight supports a rich growth of aquatic plants. The right kind and intensity of lighting will help you to recreate this luxuriant growth in your display aquarium.*

Right: *Shine a beam of white light into a glass prism and, amazingly, it emerges as a mini rainbow. The prism is revealing the spectrum of colours that make up white light. Understanding how nature uses these colours is the key to successful lighting.*

The visible spectrum ranges from violet, through blue, green, yellow, orange and red.

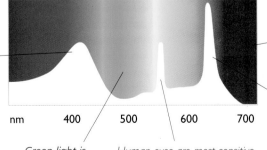

Sunlight peaks in the blue area of the spectrum. This short-wave light is used by both plants and algae.

Light in the infrared area (700-750nm) cannot be used by plants

Aquatic plants' photosynthetic ability is most sensitive to red light between 650 and 680nm.

Green light is reflected by the majority of plants.

Human eyes are most sensitive to yellow. Aquarium lights should include a peak here.

Shifting the natural light cycle

The tropical day

The typical tropical day is lit for about 12 hours from 7am to 7pm.

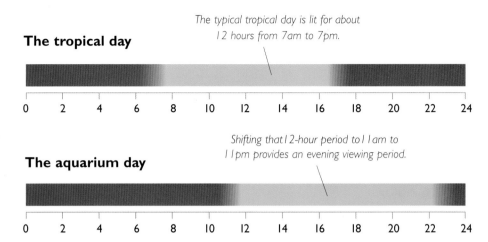

| 0 | 2 | 4 | 6 | 8 | 10 | 12 | 14 | 16 | 18 | 20 | 22 | 24 |

The aquarium day

Shifting that 12-hour period to 11am to 11pm provides an evening viewing period.

| 0 | 2 | 4 | 6 | 8 | 10 | 12 | 14 | 16 | 18 | 20 | 22 | 24 |

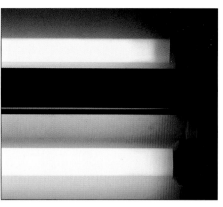

Above: *The lighting system in our featured aquarium has one triphosphor white tube and one 'pink' tube to enhance plant growth.*

end of the spectrum are less 'active' and are quickly absorbed, whereas the shorter blue wavelengths are more 'energetic' and penetrate deeper into the water before being absorbed. (This explains why undersea views in tropical diving films look so blue.) To compensate for this unequal absorption, the light source for aquarium plants should have a high peak in the red band around 650nm and also a significant peak in the blue part of the spectrum. For us to see the display, there should also be a peak in the yellow band, because human eyes are most sensitive to this part of the spectrum.

Choosing aquarium lights
Commonly used aquarium light sources fall into two categories: fluorescent tubes and metal halide (halogen) lamps. Although incandescent lights are often included in aquarium start-up kits, they are inefficient and are gradually being phased out.

Standard fluorescent tubes are a
familiar part of our domestic and working environment. They are glass tubes coated on the inside with phosphors that glow (fluoresce) when bombarded with high-energy electrons. The starter (ballast) unit triggers the process. The wavelengths of light produced depend on the phosphors in the coating. For a balanced output in

the aquarium, choose a fluorescent tube with the three required peaks of blue, red and yellow. These are best produced by triphosphor tubes – each of the three phosphors generating a different wavelength of light. For a good level of illumination, use two or more tubes above the aquarium, but this depends on how much room there is inside the hood.

Fluorescent tubes are very cheap to buy and run. A 75cm (30in) tank only needs two 25-watt tubes to illuminate it. One drawback with using fluorescent tubes is that they radiate in all directions and bounce light around the hood. This can be resolved by fitting an inexpensive reflector that directs the maximum amount of light downwards. Another drawback of standard fluorescent tubes is that their low power consumption limits their capacity to produce light bright enough for aquariums deeper than 45cm (18in).

Biax fluorescent tubes (also known as
compact fluorescent bulbs) are traditional tubes bent in two with all the fittings at one end. They are gaining in popularity. They have a higher power rating – up to 55 watts – and therefore an increased light output. They are smaller than traditional tubes, which allows more to be fitted in the same size of aquarium hood. A final advantage is that the two 'halves'

of the 'U'-shaped lamp can be coated with a different phosphors, allowing two different light 'mixes' to be emitted from a single unit.

Metal halide lamps are the best choice for deep aquariums and where really bright light is needed. These units are much more expensive than fluorescent tubes but they do produce a great deal of light, with power ratings of 70 watts plus per lamp. Most fittings have built-in reflectors that direct all the light downwards. This enables the units to illuminate deep aquariums and provide the brightness levels to satisfy even the most demanding aquarium plants. Such stark lighting can also add drama to your display, creating an impressive mix of bright focal points and pockets of deep shade. Shady corners will be welcomed by fish that prefer to spend time under cover away from the intense light. Because metal halide lamps are so powerful they also run very hot. Only use them over open-topped aquariums. This allows the lights to to keep cool but it increases evaporation from the water surface and this may cause unwanted condensation elsewhere in the home. Metal halide units are normally designed to be suspended from the ceiling above the aquarium, although some designs have side supports for them to stand on the aquarium rim.

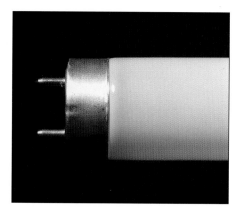

Left: The familiar fluorescent tubes used across the world in homes and offices are fine for aquarium use. The development of a triphosphor coating has boosted the output across a wide range of light wavelengths.

Left: A narrow-bore form of fluorescent tube provides intense light and saves space in the aquarium canopy. These are triggered with an electronic starter (ballast) system and are available in a range phosphor types.

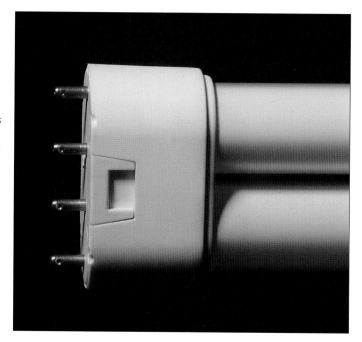

Above: The biax tube is like a conventional fluorescent tube folded back on itself with the two sets of double pins in a row of four. These tubes save space and provide high light outputs.

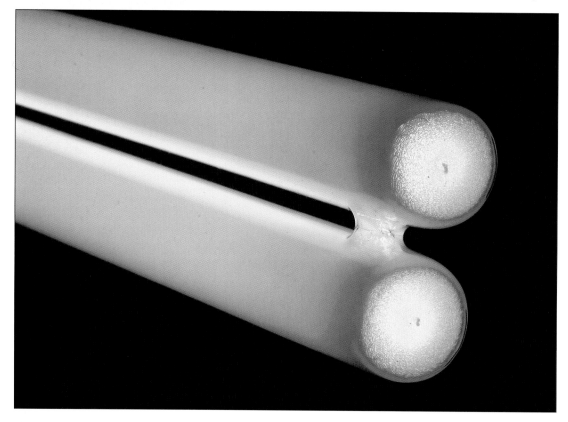

Left: Powered up, a biax tube glows just like a conventional fluorescent tube. The twin tubes close together produce an intense light for the space they occupy. Another benefit of these lamps is that the tubes can be coated with different phosphor formulations, making it possible to recreate the effect of two separate (and different) tubes in one fitting.

Above: Metal-halide lamps for aquarium use are fitted into housings such as this and suspended above the open tank. They provide very bright illumination, but must have ventilation space around them because of the heat they generate.

Right: A simple way of boosting light levels reaching the aquarium is to fit a reflector around the fluorescent tubes in the canopy or hood.

Use aquarium fittings

Always use lighting equipment specifically designed for use with aquariums, both from a safety and an efficiency standpoint. Light sources that are not designed for aquarium use will not supply the specialist needs of aquarium plants and may not be suitable for safe operation close to water.

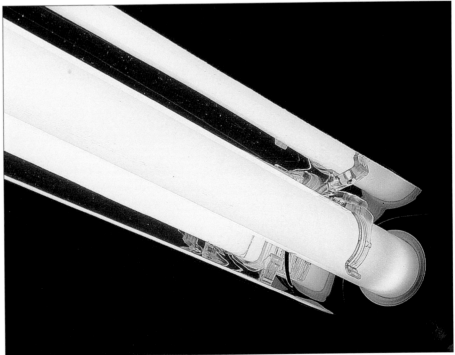

Installing the lighting system

Right: Our featured aquarium has a lighting cradle that accepts two conventional fluorescent tubes and fits snugly into the frame over the tank.

This rigid aluminium cradle will provide years of safe, corrosion-free service.

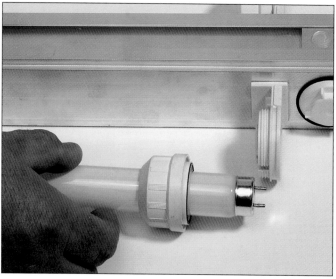

Above: Slip the locking collars onto each tube and align the pins correctly before sliding them into the sockets at both ends of the lighting cradle.

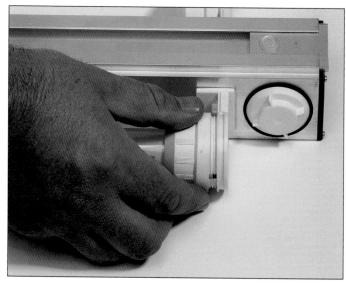

Above: Carefully tighten up the locking collars at each end to secure the tubes in place. These will make a waterproof seal around the endcaps.

Left: A view looking down into the aquarium shows the lighting cradle in place, complete with the fluorescent tubes. The cradle is part of an aluminium frame that fits over the top of the tank.

The starter (ballast) system for the tubes is housed in the central section of the cradle.

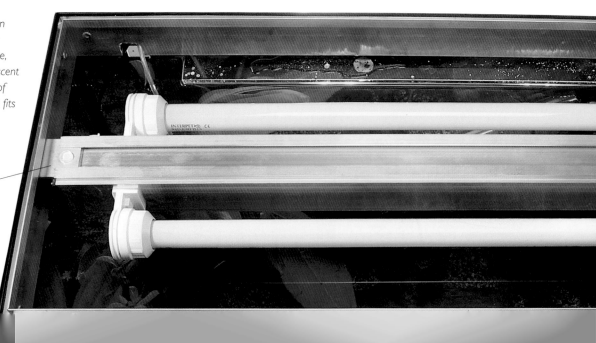

Conventional fittings

The long-established lighting method involves fitting fluorescent tubes into the hood with flying leads connecting them to the starter (ballast) system.

Above: *To protect tubes from moisture over the aquarium, waterproof endcaps are fitted on the flying leads. Align the pins to the socket and push the tube firmly but carefully into each endcap.*

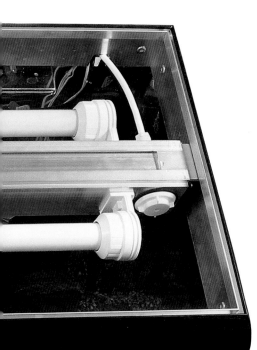

Above: *From an aquarium plant's point of view, the lighting system is the tropical sky — complete with a substitute sun providing the sole source of illumination. Plants vary in their demands for light. The aquarium may seem brightly list to our eyes, but it may be too dark for some plants to thrive.*

UNDERSTANDING WATER CHEMISTRY

Our aquarium is on the verge of becoming a miniature version of a real aquatic environment. There is a layer of fertile substrate that will nourish the aquatic plants through their first few months of life. There are rocks and pieces of bogwood to give fish shelter and security. There are lights to simulate the radiant energy of the sun. The water is warm and flowing through a filter system matched to the size of the tank. Only the fish are missing. But before we add them, it is vital for us to understand the basic properties of water and how to fine-tune them to make our display a fit environment for their continued health and vigour.

Making tapwater safe for fish

Tapwater is the most convenient source of water and is used by the majority of fishkeepers across the world. However, tapwater is supplied to our homes fit for human consumption. To achieve this, water companies treat the supply with chemicals such as chlorine and chloramine (a combination of chlorine and ammonia) to kill harmful bacteria. Even at their low dosage rate, these harsh

Above: *Tapwater is processed for human consumption. The chemicals added to it by the water company make it safe for us to drink but dangerous for aquarium fish. Treat it before use.*

chemicals pose problems for aquarium fish because they damage the delicate skin and gill membranes.

It is a 'golden rule', therefore, that all tapwater should be conditioned before use, not only when the aquarium is filled for the first time but also during water changes carried out as part of routine maintenance. Used as recommended, liquid tapwater conditioners will neutralise chlorine and chloramine and also deal with any heavy metals that the water has picked up during its journey to your home. The most advanced products also contain an aloe vera colloid to provide a protective layer in addition to the fishes' own mucus coating and add beneficial bacteria to boost the natural biological cycles that keep the water clean in a well-balanced system.

For large aquariums (over 1.8m / 6ft long) investing in a reverse-osmosis (R.O.) water purifier may be a better long-term option. These units contain a semi-permeable membrane that effectively 'sieves out' dissolved substances at the molecular level. In fact, the resulting water is so pure that minerals need to be added back into it to make it suitable for fishkeeping. Getting the right balance of minerals can be difficult for beginners. If buying an R.O. unit does not make economic sense, you can buy supplies of R.O. water from aquatic retailers.

What does pH value mean?

The pH value is an indication of whether water is acidic, neutral or alkaline. The scale extends from 0 to 14, with 0 being extremely acidic and 14 extremely alkaline; pH 7 represents the neutral point. It is important to realise that the pH scale is logarithmic, with each unit reflecting a tenfold difference. Therefore, pH 6 is ten times more acidic than pH 7 and 100 times more acidic than pH 8.

The pH of natural water supplies can be 'shaped' by various influences, including the amount of pollution in the air through

Above: *Treating tapwater with a conditioner is a simple process. Following the maker's instructions, add the required amount of conditioner to tapwater in a bucket before use in the tank.*

Below: *A reverse osmosis unit will filter out suspended particles and remove all dissolved substances at the molecular level. R.O. water is so pure it need added salts for fishkeeping use.*

which rain falls, the surface soil on which it lands and the underlying rock over which it flows. Once inside the aquarium, other factors continually affect the pH of water, such as biological filtration and the respiration of plants and fish.

It is essential to check the initial pH of water intended for the aquarium and to continue checking it in order to maintain a uniform pH value. Fortunately, it is easy to determine the pH value of water by using a test kit, which can be a dip strip, tablet or liquid test kit. Whichever method you choose, be sure to pick one that is easy to use and is packed with clear information about the results you get.

A good aquatic retailer will be able to tell you whether your local water supply is suitable for the species of fish you wish to keep or whether it is necessary to adjust the pH value in order for the fish to thrive in your system.

Changing the pH value of your aquarium water is easy to do with a pH adjuster. Follow the manufacturer's instructions carefully, as changing pH should be carried out at the maximum rate of 0.3 of a pH unit per day when fish

The pH scale

The two elements (hydrogen and oxygen) that make up water exist as positively charged hydrogen ions (H⁺) and negatively charged hydroxyl ions (OH⁻). The pH scale reflects the relative amounts of these two ions. Water with more hydrogen ions than hydroxyl ions is acidic, whereas a greater number of hydroxyl ions makes the water alkaline. Water with equal numbers is neutral.

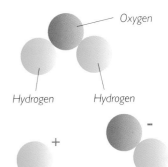

Water molecule (H₂O)

Oxygen

Hydrogen   Hydrogen

Hydrogen ion (H⁺)   Hydroxyl ion (OH⁻)

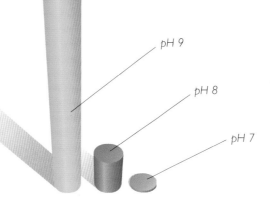

pH 9

pH 8

pH 7

The pH scale is logarithmic, meaning that each unit change in pH, say from 7 to 8, is a ten times change; from 7 to 9 is a hundred times change, and from 7 to 10 reflects a thousand times change. This is why sudden changes are stressful to fish.

Contrasting pH values

The natural environments of commonly kept aquarium fish can differ quite markedly in the pH value of their native waters. Discus *(Symphysodon discus)* live in acidic Amazon tributaries in South America at pH 6, whereas cichlids in Lake Malawi in eastern

Africa thrive in alkaline water at pH 8-8.5 . Since the water of the Amazon environment is one hundred times more acidic than Lake Malawi it is not surprising that you never find these two types of fish happily thriving in the same aquarium.

Lake Malawi cichlids live in hard, highly alkaline waters.

Discus come from areas with soft, acidic water.

93

are in the aquarium. Ideally, change the pH before introducing fish. Also be aware of the pH value of the water the fish are coming from before introducing them to your aquarium, with a maximum of 0.3 pH difference between the two.

Hard and soft water

As the rock over which water runs helps to determine its pH value, it also influences its hardness. Water hardness is a measure of the dissolved salts it carries, mainly carbonates and sulphates of calcium and magnesium. Water with a high salt content is referred to as hard, while water low in these salts is referred to as soft.

The total (or general) water hardness value is measured in degrees of hardness (°dH) and can be divided into two types: temporary hardness, which can be removed by boiling the water; and permanent hardness, which cannot. Test

kits are available for both total and temporary hardness. The scale for total hardness ranges from 0°dH (very soft) to 28°dH and over (very hard).

A good test kit will advise you on the ideal fish to keep for a particular hardness and provide practical notes on changing the hardness if required. Increasing water hardness is fairly straightforward and involves adding balanced salts to the aquarium. Making water softer used to be a difficult process, but with the advent of liquid water softeners and reverse osmosis units, reducing water hardness is relatively easy.

In most natural environments, the pH value and hardness are linked, with low pH water being soft and high pH water being hard. The hardness of your aquarium water will also give you an indication of how stable the pH is likely to be. In soft water, the low level of dissolved mineral salts provides less

resistance (buffering capacity) to changes in pH, which tends to drop. Hard water has a greater mineral content and is said to have a high buffering capacity, and so the pH remains more stable.

Fish and plants prefer differing hardness values if they are to thrive. Your local aquatic retailer will be able to guide you towards fish species suited to your local water conditions or advise you on changing your water hardness and/or pH value to enable you to keep the fish you have set your heart on.

The nitrogen cycle

Understanding the nitrogen cycle is the single most important aspect of successful fishkeeping. It is a natural process that occurs in every aquatic habitat around the world. The key compounds in the cycle all contain nitrogen (N_2), which makes up 80% of the air we breathe and on its own is a 'harmless' inert gas.

The nitrogen cycle

Your aquarium is a miniature version of the natural world. Powered by bacteria, the nitrogen cycle converts nitrogen-rich wastes into a series of successively less poisonous substances.

Nitrates are readily taken up by plants as a food source, usefully reducing the levels in the tank.

Nitrobacter bacteria convert nitrite to nitrate (NO_3), which is much less harmful to the aquarium fish.

After digesting nitrogen-rich foods (such as protein), fish excrete extremely toxic ammonia (NH_3) directly from the gills and also in urine and faeces. Decomposing biological matter, such as uneaten food and dead leaves, add to the ammonia load in the aquarium.

Nitrosomonas bacteria in the biological filter (and also coating surfaces within the aquarium) convert ammonia to nitrite (NO_2), which is still dangerous, even at low concentrations.

Upsetting the nitrogen cycle

Chemical treatments added to the aquarium will reduce the level of nitrifying bacteria in your filter. Always monitor the water quality during and after their use to ensure that no undue stress is placed on the fish.

Household cleaners used near the aquarium can cause a bacterial 'wipeout' in the filter. Be careful using furniture polish or glass cleaners near the tank and never use them on the hood, where the risk of contamination is at its greatest.

If you detect an unexplained rise in ammonia and nitrite, check the aquarium carefully for dead fish and/or uneaten food as these can cause a dramatic deterioration in water quality.

Combined with hydrogen, however, nitrogen makes ammonia (NH_3), an extremely poisonous substance. Even at very low concentrations, it will rapidly kill all the fish in your aquarium. This is a convenient, if dramatic, point to break into the nitrogen cycle and trace how ammonia is converted into progressively less toxic compounds until the cycle starts all over again.

Life can poison itself

All plants and animals in the natural world respire, using oxygen to release energy from food (which includes nitrogen-rich amino acids and proteins) and produce organic waste. Plants produce organic waste relatively slowly through the gradual decomposition of their leaves and stems. Aquarium fish produce organic waste much more rapidly, actively excreting ammonia from their gills as a direct byproduct of respiration. They also produce organic waste in the form of small amounts of urine and also solid faeces. All this waste forms free ammonia in the water – potentially a self-poisoning situation for the creatures living in it.

Free and ionised ammonia

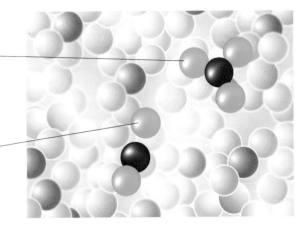

Free ammonia (NH_3) consists of one atom of nitrogen and three of hydrogen.

The nitrogen atom of ammonia attracts a hydrogen atom from water (H_2O).

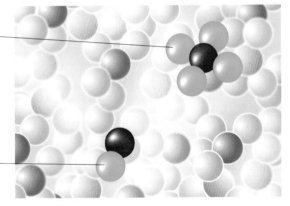

The hydrogen atom attaches to the nitrogen atom, conferring a positive charge on the newly created ammonium ion (NH_4^+).

Without one hydrogen atom, the water molecule becomes a negatively charged hydroxyl ion (OH^-).

Bacteria to the rescue

Of course, nature would not allow life to poison itself. As in all natural cycles, what is waste for one set of 'creatures' becomes food for another. The rescue party in the aquarium (and all aquatic habitats) is led by bacteria that consume the ammonia and use oxygen to convert it into nitrite (NO_2) – a process called nitrification. These aerobic (oxygen-loving) bacteria are called *Nitrosomonas*.

The bad news is that although nitrite is less toxic than ammonia, it is still deadly in relatively small quantities. Fortunately, a second set of aerobic nitrifying bacteria, called *Nitrobacter*, use oxygen to convert nitrite into nitrate (NO_3), which is relatively harmless.

The cycle turns

The production of nitrate provides the impetus to keep the cycle turning. Nitrate is used by plants as food for growth and the nitrogen it contains becomes incorporated into the amino acids and proteins that make up the structure of leaves and stems. When the plants die and decompose or are eaten by fish, the waste produced releases ammonia into the water and the nitrogen cycle takes another turn.

How toxic is toxic?

'Free ammonia' (NH_3) is extremely toxic, easily passing from the bloodstream into the tissues and brain and causing damage and behavioural impairment. Levels over 0.02mg/litre (equivalent to 0.02ppm – parts per million) are poisonous to fish. For any given amount of ammonia in the water, the proportion in the free form rises as the temperature and/or pH value increase. In cooler more acidic water, more of the ammonia is in the less harmful ionised form (NH_4^+, ammonium). A 'golden rule' for successful fishkeeping is a zero reading for ammonia levels.

Nitrite combines with the oxygen-carrying pigment haemoglobin in the fish's blood, preventing the red blood cells from picking up oxygen. (Affected blood turns from a 'healthy' red to a 'damaged' brown colour.) A classic sign of nitrite poisoning is fish gasping at the surface. This is not a lack of oxygen in the water, simply that the fish are suffocating as their blood cannot carry enough oxygen. If you see this, always take immediate action to reduce nitrite levels, such as making a water change (see page 150). Lethal nitrite levels differ between fish species, but 0.2mg/litre can be fatally toxic. Again, the only acceptable level in the aquarium is a zero reading.

Nitrate is relatively harmless in comparison to ammonia and nitrite. High nitrate levels can cause health problems for young fish and fry; however, the main problem will come in the form of unwanted algae growing the aquarium.

Nitrate is a major nutrient for plants and any excess will fuel a growth explosion in a wide range of algae species that will rapidly make a mess of your aquarium. It is important to monitor nitrate levels and keep them below 40mg/litre. A healthy population of aquarium plants can keep levels of nitrate below 10mg/litre.

It is always worth testing your tapwater for nitrate, since filling the aquarium from the tap will introduce a certain level to your system. This nitrate comes from fertilisers used in agriculture that wash into the water supply chain and are eventually piped to your house.

How to test water

Controlling the water conditions is the key to a successful aquarium. A vital part of this is to measure, record and compare the values of a number of characteristics that reflect the state of the water circulating around the tank.. Here, we look at the practical details of testing water for all the major parameters, from pH to nitrate levels.

Using test kits properly

The test kits available today are easy to use and provide accurate results. The most popular kits involve adding liquid or tablet reagents to a water sample and comparing a colour change to a printed chart. 'Instant-result' dip strip tests are also available for many parameters. Tablet kits are probably the best ones for consistent results because the amount of reagent is provided in a precisely measured dose.

If you are taking a water sample in a test vial provided in the kit, rinse it first in the water you are about to test, i.e. the aquarium water. Also, when the filling the vial to the desired level (usually 5ml or 10ml) ensure that the bottom of the meniscus is on the 5 or 10ml mark. This will ensure that the test is as accurate as possible. Hold liquid reagent bottles fully inverted to ensure that a uniform drop size is achieved from the dropper. Varying drop sizes will influence the results and not give you an accurate reading.

Once you have finished your tests always wash out and dry your test vials

Left: When taking a water sample, select a clean test vial from the kit and rinse it first in the aquarium. Washing it in tapwater may distort the final test results.

Below: Tests usually involve starting with a water sample of 5 or 10ml. For accurate results, make sure the bottom of the meniscus sits on the line printed on the vial.

Above: Adding the right amount reagent is also vital. Invert bottles for standard-sized drops. Tablets contain a measured dose.

Keeping a record

During the first few weeks of setting up the aquarium it is worth keeping a note of the date and the results of your tests. This could be invaluable information should you have problems in the future that you want to discuss with your retailer.

pH test

The pH test indicates how acid or alkaline the water is. Tests are available to provide broad-range (shown here) and narrow-range readings. It is a good idea to continue testing the pH throughout the life of the aquarium because ongoing processes such as biological filtration will affect the pH value of the water over time.

Water hardness test

Hardness test kits should be able to register both temporary and total hardness to give you the full picture of your water chemistry. Most are simple colour change kits in which the number of drops required to change the colour of a sample indicates the degrees of hardness.

Ammonia, Nitrite, Nitrate

When testing for any of these substances it is best to carry out tests for all of them because this will reveal the complete biological status of the aquarium. If you simply test for ammonia and you are happy that you record a zero reading, you will probably have a high level of nitrite and your fish will be suffering. Once the first fish are added you should test the water every day until both ammonia and nitrite are under control. Then repeat the tests weekly to ensure that the biological balance of the aquarium is continued with the addition of any new fish. Repeat the tests for a week whenever new fish are added to the aquarium.

Right: This total ammonia test shows a reading of 0.4mg/litre (0.4ppm - parts per million). This toxic short-term level will fall as filter bacteria respond.

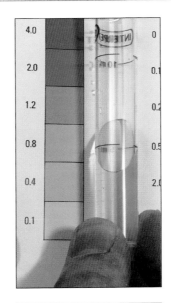

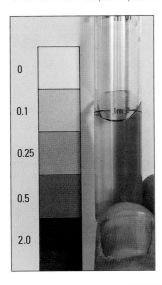

Left: A nitrite result of 0.25mg/litre can fatal for some fish species and needs to be reduced to zero for safe fishkeeping.

Right: Nitrate levels can go quite high, especially in sparsely planted tanks where the plants do use enough of it as food. This 25mg/litre result is typical and safe in most aquariums.

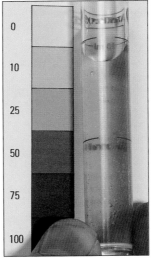

Dip tests

Dip tests consist of paper or plastic strips with impregnated pads that change colour when the strip is dipped briefly in a water sample. Comparing the colour change against a printed reference chart provides readings of several parameters. While one-off tests are a good introduction to water testing they are not the best option for long term use as the risk of contamination of the test strips is quite high.

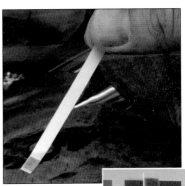

Above: Dip the test strip directly into the aquarium.

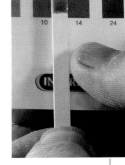

Right: Compare the colour with the printed chart. This test shows a water hardness of 14°dH.

Phosphate

Testing for phosphate is usually recommended if the aquarium appears to be suffering from an algae problem. Some phosphate may come in with the tapwater used to fill the aquarium and is also an ingredient in fish food. If allowed to rise above the level at which the aquarium plants can use it as a fertiliser, the excess will become food for fast-growing algae. A phosphate test is shown in action on page 146.

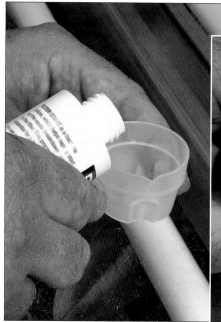

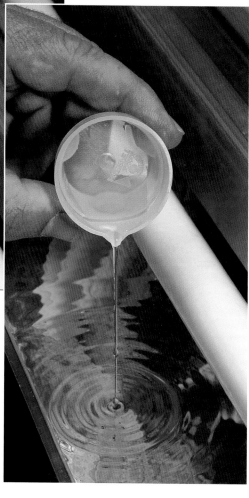

Above: *Carefully measure out the correct amount of filter start-up product needed for your aquarium system. Be sure to allow the precise instructions provided with the product.*

Right: *Add the measured amount directly to the aquarium water. These products provide heterotrophic bacteria that will foster the growth of beneficial bacteria to deal with nitrogenous waste.*

How your biofilter works

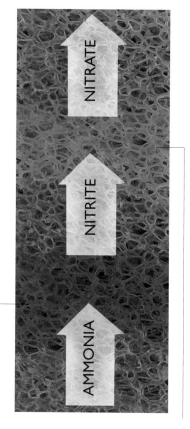

Nitrosomonas bacteria grow in the foam and convert ammonia into nitrite.

As levels of nitrite increase, Nitrobacter bacteria convert it into the less toxic nitrate.

before storing the whole kit in a safe place away from children, as many reagents used in water tests are potentially harmful. Keep test kits out of direct sunlight, not only to keep the chemicals stable but also to prevent colour reference charts fading in bright light. Refills are available once you have used all the reagents, which saves you having to buy the entire kit again.

The water tests featured on pages 96-97 are the most important ones used in the aquarium hobby.

Maturing the filter

As we have seen, the nitrogen cycle is nature's way of dealing with its own waste products. However, in the highly managed aquarium environment we must harness the cleansing power of the nitrogen cycle bacteria effectively so that they can deal with the waste produced by the livestock.

How your filter works

The nitrifying bacteria that break down nitrogenous waste will coat all surfaces inside the aquarium but not in sufficient numbers to control the amount of waste produced by your fish. This is why we need a biological filter. The filter medium, in our case the blue filter foam, provides a huge surface area in a compact space on which bacteria can grow. Eventually, the bacteria will multiply to fill the filter sponge and control the waste produced by the fish, giving zero readings on

ammonia and nitrite test kits. However, this can take 6-8 weeks, a long time when we want to enjoy our display and stock it with fish.

The key point is that the *Nitrosomonas* and *Nitrobacter* bacteria thrive best growing within a colony of other species, the so-called heterotrophic bacteria. Providing these will cut filter maturing time in half. Heterotrophic bacteria are available in liquid or powder form in filter start-up products from aquarium shops. Simply add one of these to the aquarium on a regular basis during the first few weeks, beginning as soon as the filter is switched on. Even if the aquarium only has plants in it, there will be a small production of nitrogenous waste to help start the biological filter. Be sure to follow the specific product manufacturer's instructions when using these materials.

Once you have started to add a filter start-up product you need to monitor what is happening in the filter by using aquarium test kits. Testing for the levels of ammonia, nitrite and nitrate in the aquarium water will show how the bacterial growth is progressing. Take the tests every two days for the first month to six weeks and make a note of the readings. This provides an ongoing picture of what is happening in the aquarium.

Within a maturing filter the bacteria will only multiply enough to consume the amount of waste present, and no more. (This means that every time new fish are added, the bacterial colony will increase to cope with the extra biological load.) The first signs of anything happening will be detected a few days after introducing the first fish. Although the start-up product will have added the necessary family of bacteria to the filter, not enough of the *Nitrosomonas* bacteria will have grown to cope with the amount of waste produced. Water tests will show a rise in the ammonia level created by the waste from the fish. Even minimally fed fish produce ammonia as a continual byproduct of respiration. If not dealt with, the ammonia will climb to lethal levels in a few days, so it is vital to control this.

Clear water may be deadly

My water is crystal clear, so it must be OK. Why do I have to use a test kit? Every aquatic retailer has heard this question a million times. The simple answer is that the deadly toxins ammonia and nitrite are invisible. The water may be gin clear but it could be deadly and kill fish in minutes. The only way to be sure is to check the water quality using a test kit. These are simple to use and for a minimal investment can save the lives of the fish and other creatures in your care.

Controlling ammonia

The traditional way of controlling ammonia has always been to carry out a water change – simply removing ammonia-laden water from the aquarium and replacing it with fresh water that contains no ammonia. While this is good for the fish, it starves the filter bacteria of their ammonia 'food' and thus stops the colonies growing larger. In emergencies, water changes of up to 70% can be carried out. (If you do this, make sure that the conditioned water you put back into the aquarium is at the same temperature and pH as the water removed.)

The best strategy is to use an ammonia-removing product, ideally in a convenient liquid form. Choose one that will detoxify the ammonia and still allow the bacteria to 'feed' on the nitrogen part of the treated molecules. Use a test kit to keep track of the reducing ammonia levels to reassure you that the treatment is working.

In addition to ammonia-detoxing liquids, there is an ammonia-removing mineral called zeolite. This white granular substance can be installed in the filter compartment and will remove vast quantities of ammonia like a water change. It will starve the maturing bacteria but provides a good emergency option.

You may find ammonia readings on your test kit for a couple of weeks. These will reduce at the same time as nitrite readings start to increase. This is telling us that the *Nitrosomonas* bacteria have multiplied enough to deal with the amount of ammonia being produced by the fish and that nitrite is being produced as a result.

The *Nitrobacter* bacteria that 'process' nitrite into nitrate grow alongside the *Nitrosomonas* bacteria in the filter media. However, their growth rate is hampered by the presence of ammonia. This means that nitrite levels can get quite high unless remedial action is taken while the *Nitrosomonas* break down the ammonia.

Below: Once an aquarium becomes established with a balanced population of fish and plants, the biological filter will keep the water clean and healthy, but regular water tests are still vital.

How a biological filter matures in the first few weeks

This graph shows the typical rise and fall of ammonia, nitrite and nitrate levels as a biological filter matures. It reflects the sequence of events in our featured aquarium.

Above: *More fish can be added as the biological filter matures.*

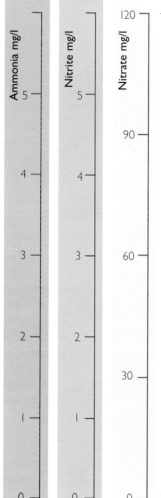

Ammonia mg/l

Nitrite mg/l

Nitrate mg/l

120

90

60

30

0

5

4

3

2

1

0

— *To keep the comparable readings visible on the page, the scale for nitrate is different from those for ammonia and nitrite.*

First fish added

Looking at the graph, you will notice that the amounts of ammonia, nitrite and nitrate peak one after the other as beneficial bacteria in the biological filter convert one substance to the next.

Second fish added

Ammonia levels rise just after adding new fish to the aquarium.

Water change

Water change

Water change

Water change

WEEK 1 / 2	WEEK 3	WEEK 4	WEEK 5	WEEK 6
During the first two weeks a filter start-up product has been added to help the filter bacteria mature. This product is purely bacterial and does not introduce any ammonia to the system. This enables fish to be added at the start of week three.	*Fish are introduced. These produce ammonia directly and through waste products. The bacteria already present start to convert this into nitrite and nitrate. More bacteria will colonise the filter media to cope with the increasing ammonia levels.*	*The waste from the new fish is now being processed by the filtration bacteria. Keep an eye on both ammonia and nitrite levels during this time to ensure that the fish are not exposed to dangerous levels of these toxic wastes.*	*Now a few more fish can be added. The already established filter bacteria can cope much better with the excess ammonia this causes. The growing level of nitrate is reduced at the end of each week by a 25% water change.*	*The filter is approaching maturity. Waste produced by the first two batches of fish is now easily controlled by the ever-increasing bacteria. The filter start-up product has provided the right environment for these bacteria to thrive.*

The ideal target for ammonia and nitrite levels is zero. Small blips arise when new fish are added, but these are readily reduced by the filter bacteria.

Water change

Third fish added

Water change

When to use filter start-up products

Filter start-up products should be used with a new setup and whenever the biological loading on the filter changes, such as when new fish are added or whenever test kit readings show that the bacterial filter population is not coping with the amount of waste in the system. This may take place when medications and other tank treatments have been used or shortly after a filter has been cleaned.

Nitrate levels can be reduced by water changes and any excess will be absorbed by healthy growing plants as a food source. It is important o keep checking nitrate levels to avoid an algae bloom in the aquarium.

Controlling nitrite

There are no chemical treatments to remove nitrite. Water changes are a simple way of reducing the level and in extreme cases aquarium salt and methylene blue can be added to 'mop up' excess nitrite, although it is difficult to know how much to use.

Once the ammonia level falls to zero, the *Nitrobacter* population will increase to the level at which these bacteria can deal with the volume of nitrite being produced. Nitrite readings can decrease rapidly, sometimes overnight, to zero.

Controlling nitrate

Once your test kits show zero nitrite readings you will see nitrate levels rise. This may not occur in heavily planted tanks because the plants are using all the nitrate as food. However, if nitrate levels do rise, consider making a water change to dilute the concentration. Excess nitrate will trigger algae problems in the aquarium. Always test your tapwater for nitrate, as it may already be carrying substantial quantities. Passing the water through a nitrate-removing resin (shown on page 25) is effective and harmless.

How long does it take?

The maturation process will take a few weeks, although every tank is different. The numbers and size of the fish, the amount of food introduced, and the size and turnover of the filter will all affect the process. Always remember that every time the biological loading changes, the filter will take time to respond. Even a mature filter will need to increase its bacterial population when you add new fish to the aquarium. Water testing is vital to ensure that any blips in water quality do not endanger your livestock.

Once your filter has sufficient bacteria to deal with the waste produced, your test kits will show zero readings for ammonia and nitrite, and little or no reading for nitrate. At this point, your filter is working at full efficiency and you have control of the nitrogen cycle.

WEEK 7	WEEK 8

Adding fish has hardly any effect on the levels of ammonia and nitrite. This is because the bacteria in the filter have reached a population that can quickly adapt to any changes. This shows that the aquarium is healthy and balanced.

The biological filter is fully established and the aquarium is starting to mature. More sensitive fish can now be introduced. Check nitrate levels; you may need to use a chemical filter medium to reduce them.

New developments

Ongoing developments in biochemistry have resulted in start-up products containing live bacteria that may eliminate the maturing process altogether. These need to be kept fresh and supplied with ammonia in the shop before you buy them. When it becomes possible to create a stable form of preserved Nitrosomonas and Nitrobacter bacteria, this will represent a breakthrough for the fishkeeping hobby.

BRINGING THE AQUARIUM TO LIFE

Three weeks after the initial setting up, the plants are growing and the different areas of the display are starting to blend together. Do not worry if some species do not appear to have developed much; growth rates vary and in time all the plants will flourish as they settle into their new home. To keep the aquarium in pristine condition, clean and without any evidence of algae or other problems, it is important to begin a maintenance regime at this stage. This is a habit you must acquire, but usually only involves carrying out small but regular tasks that are all part of being a successful aquarist and, hopefully, more of a pleasure than a chore.

Be sure to turn off the power supply to the aquarium before putting your arms into the water. If the tank is very full, first remove up to a bucketful of water, otherwise the water will flood, which is both messy and potentially very hazardous. The third week also begins the most exciting aspect of creating your display: bringing the aquarium to life by introducing the first fish. However tempting it may be, do not buy all your fish at once; follow the guidelines set out on page 108.

With the aquarium planted up at the end of day two (above), it was left to settle. Since then, the plant roots have established and there is clear evidence of growth (right).

Plants have given our living picture a framework that will come alive with the addition of some well-chosen fish.

TIDYING UP THE TANK

Plant growth is evidence of life in the aquarium, but this means that smaller life forms, namely algae, are also trying to gain a foothold. If you inspect the front glass, you will probably notice the appearance of small green spots, both here and on the rocks and internal equipment. This spot algae can be removed with a wad of filter wool, which is effective even on the most stubborn encrusted algae growths. The algae scraper and the magnetic algae cleaner are more up-to-date tools for this job. In good models of algae scraper, the handle on the scouring pad should be slightly angled to enable you to access the corners of the tank and exert more pressure when required. To use a magnetic algae cleaner, place the magnet faced with the scouring material inside the tank and guide it from outside with the other magnet. The internal magnet has a string attached so that you can retrieve it from the base of the tank if it sinks. Modern magnets with internal air pockets will float if they become detached.

Trimming plants

Some plants will be growing vigorously and may need trimming. The stems of tall plants, such as hygrophila, often grow out in the wrong direction and if left, would ruin the display. These can be simply trimmed back using a pair of sharp scissors. Always cut just above a node – the point where the leaves are attached to the stem – to stimulate new growth from the buds at the base of each leaf. With luck, these will grow out in the right direction.

Above: An algae magnet enables you to clean the glass inside the aquarium easily without getting your hands wet.

Above: A wad of filter wool effectively removes algae and debris from the interior glass surface. You can clearly see the dirt accumulating on it. Dispose of the wool when you have finished.

Right: Using a scraper on the end of a handle is another method of cleaning the glass. The rough surface deals safely with stubborn marks. Clean glass is a sign of a well-cared-for aquarium.

Some foreground plants can grow quite tall and will need regular trimming to keep them under control. This is the case with the hemianthus in the featured aquarium. The aim is to ensure that it stays bushy and does not grow leggy or obscure the swimming space behind it.

Now is also the time to attend to the larger-leaved plants towards the back of the display. Using scissors, remove any damaged, algae-encrusted or generally spent leaves. Trim these as close to the base of the plant as possible, without damaging any new growth.

Right: The hemianthus in the foreground at the lefthand side of the aquarium has produced some long shoots. To encourage a bushy habit, trim away the excess growth with sharp scissors.

Left: *The hygrophila also needs trimming. Think about where to snip off a long shoot. Cutting just above a node will encourage the plant to produce new, bushy growth.*

Below: *This long shoot is growing out sideways and spoiling the appearance of the display, so cut it off. You can replant these trimmings to fill a gap elsewhere.*

Coping with larger plants

Left: When removing foliage from a large-leaved plant such as this echinodorus, cut the stalk at a point near the base. This encourages new leaves to develop and is a good way of maintaining compact growth.

Below: Algae is growing on this leaf, so dispose of it. Older, larger leaves generally die back first, which is a shame as these are the ones that make the most impact, initially, but new ones will soon take their place.

Right: It is interesting to compare this view of the tank before the glass was cleaned and the plants trimmed, with the view below. If you neglect to monitor and trim the plants, stronger varieties can soon swamp smaller, less vigorous ones.

Below: *Do not be afraid to trim back wayward shoots or damaged leaves, even if it means exposing some of the wood, rocks and equipment again. Healthy plants grow back rapidly.*

CHOOSING AND ADDING FIRST FISH

With the aquarium up and running reliably and the plants in place, you can think about adding the first fish. Bear in mind that you are introducing them into an environment where very little biological activity is taking place. The natural cycles that will sustain the fish only start to mature once the fish have been added, although you can accelerate the process by adding a filter start product as explained on page 98 .

Nevertheless, there is always a chance that the first fish will be exposed to less-than-perfect water conditions during the maturation process, so it is important to select species that are known to be hardy. This means that they are able to tolerate changes in water quality during the maturation process. It is not an excuse for allowing the water quality to deteriorate, as any fish will suffer in poor conditions. Indeed, if anything, you should tend the system with extra care in the early days to ensure that the water quality remains as good as possible for the first fish.

The first fish

While the choice of 'first fish' is open to debate, and everyone has their own favourites, the fish featured on pages 112-115 are good candidates for consideration. Always bear in mind that even though their hardiness is one of the principal reasons for choosing them, first fish will form part of the finished display and you need to consider how they will coexist with the species you decide to add later on. As happens in nature, not all living creatures will live happily together. Always ask if the fish you want to buy are suitable for a community aquarium. This avoids the risk of buying a beautiful fish that eats all the other occupants of the tank within 24 hours!

Buying fish

You can buy the equipment for your aquarium from anywhere, but when it comes to buying the fish, you must be satisfied that they are coming from an

How many fish should I buy the first time?

As most of the first species added to the featured aquarium are small shoaling fish, it would be wise to start with a maximum total fish length (excluding the tail) of 25cm (10in) for an aquarium measuring 60-90cm (24-36in) long. A 120cm (48in) tank could accommodate up to 38cm (15in) of total fish length as an initial stocking. These total fish lengths can be made up of more smaller or fewer larger individuals.

Below: *Shoaling fish, such as these zebra danios, settle down into a new aquarium more quickly if introduced as a group.*

Left: *Taping the corners of the bag ensures that there is no risk of small fish becoming trapped. The top of the bag should be securely knotted. Hold it firmly while transporting the fish.*

Below: *Following a short journey home, float the unopened bags in the aquarium for about 20 minutes to allow the water temperatures to equalise before releasing the fish.*

Above: If the journey home was long, first roll down the sides of the bag into a collar to allow stale air to escape. Some fishkeepers turn off the aquarium lights at this stage to avoid stressing the fish.

Although this is a potentially stressful time for fish, they should recover quite quickly.

Below: Float the bags in the aquarium for 20 minutes or so. Check whether the water temperatures have equalised by dipping your finger first into the bag and then the tank.

outlet that sells good-quality, really healthy stock. It is worth visiting all your local retailers to compare the quality and price of fish. As a guide, consider the following points when choosing a retailer:

- The retailer should have a licence to sell live animals.
- The retailer should be a member of an aquatic trade association with rules on keeping its livestock. This is evidence of a professional business that takes its responsibilities seriously.
- Is the shop clean? Are all the aquariums clean, inside and out? This shows a high degree of care by the shop.
- There should be no dead or diseased fish on show. Do not buy any apparently healthy fish from an aquarium with diseased fish in it. The chances are that all the fish will be

affected. Ideally, fish in quarantine should not be on view.
- All the tanks should have labels detailing the fish species they contain, their price, potential size, any special food requirements, compatibility with other fish and any special requirement, such as 'likes plenty of plant cover'. The best shops will also indicate the date on which the fish arrived on the premises.
- Look at the fish; do they look bright, healthy and interested in their surroundings? Are they behaving normally?
- Are the fish guaranteed? Most shops will offer some kind of guarantee, but

once the fish leave the shop you are entirely responsible for them and their environment. Any guarantee may involve testing the water in your aquarium should there be a problem.
- If possible, watch another customer buying fish and observe the staff. They should handle the fish as little as possible to minimise stress.

Taking your fish home
Your retailer will transfer your chosen fish to one or more plastic bags, with enough water to cover them, and then fill the rest of the bag with air before sealing it. If the journey home is less than an hour, the fish

Above: *Hold the bag just below the water surface, tipping it slightly to encourage the fish to swim out in their own time. Make sure they are all released. Repeat the process with each bag of fish.*

will be fine with just air in the bag. If it is any longer, the retailer should fill the bag with oxygen. Finally, the retailer should tape the corners of the bag to ensure that no fish become trapped. For safe transportation, it is a good idea to take a cardboard box with you to the shop to contain the bags on the journey home.

Once you have found a good retailer you can choose your first fish with confidence. They will bring life and movement to the backdrop provided by the plants and become the real 'stars' of the living picture you have created. All the first fish are top to midwater swimmers so will add instant interest.

Above: These three pentazona barbs are soon swimming around the aquarium, investigating their new surroundings. Some fish are more timid and may hide away at first, but this is quite natural.

Left: With all the starter fish installed, the tank is now home to five zebra danios, five harlequin rasboras and three pentazona barbs. Do not be surprised if they do not show their best colours straightaway. These will improve as the fish become more confident in their new surroundings.

Stocking levels

As a rule of thumb, work on the basis of 1cm (0.4in) of fish (excluding the tail) for every one litre of water. This means that to an aquarium with a water volume of 150 litres you can add a maximum of 150cm (60in) of fish. This may seem quite high, but the calculation is based on the maximum adult size of the fish. This is why it is important to be aware of how big your fish will grow, and to build up the stock gradually. For example, zebra danios can reach a maximum size of 5cm (2in). If you add six individuals to the tank, you will have accounted for 30cm of ultimate fish length, even though each young fish only measures 2cm (0.8in) when it first goes into the aquarium.

As a guide, never add more than 40cm (16in) of adult size fish at any one time. This will allow the biological filter to mature so that it can process their waste. Monitor the water quality until both the ammonia and nitrite levels have again reduced to zero before adding more fish. This is the only way to be sure that your filter has matured sufficiently to process the waste produced by all the fish in the aquarium.

FIRST FISH FOR YOUR AQUARIUM

The first fish you add to the aquarium immediately bring colour, movement and drama to the display. They are all midwater or upper-layer swimmers and will add interest to an area of the display perhaps not yet reached by the growing plants. The species described here include danios, rasboras, tetras and barbs and are hardier than most, but still require care and attention. It does not mean that all the fish in these groups are suitable for consideration as 'first' fish. For example, certain tetras should be added only when the tank has been running for several weeks, if not longer. Other representatives of these groups of fish will be examined in more detail later in the

book, as 'second' or 'third' fish. Just as you planned the decor and planting of your aquarium, it is worth thinking ahead about the fish you may wish to add in the future. Looking at the second and third fish options, and deciding which ones you like, may help you decide which species to start with, as they will all need to fit in together in the final display. For this reason, it also pays to discuss your future choices with your retailer when you buy the first fish. At this time, remember to include food and a net if you do not have these items already. Finally, follow the buying guidelines on page 108 and do not rush into buying fish you suspect are not healthy.

Leopard Danio
Brachydanio rerio var. frankei
This variety of the zebra danio has spots along its flanks instead of stripes. It will mix happily with zebra danios and has the same temperament as its close cousin. In common with all danios, it has a pair of barbels on each side of its mouth that it uses to find food. This is simply a question of do you prefer spots or stripes?

Zebra Danio
Brachydanio rerio
As its name implies, this danio has distinctive horizontal stripes along each flank. Zebra danios are very lively schooling fish that swim at all levels and prefer a well-planted aquarium in which they can dart from cover to cover. They will thrive best in a school of six or more and are a popular choice for community aquariums. Originating from eastern India, they prefer neutral to slightly acid water conditions, although captive-bred specimens will tolerate a much wider pH range. Zebra danios are also available in a long-finned variety. Maximum adult size: 6cm (2.4in).

Black Widow Tetra

Gymnocorymbus ternetzi

The black widow tetra is an excellent starter fish. Originating from South America, this peaceful schooling fish appreciates the cover of bushy plants and the shade generated by tall species. Younger fish can be identified by their striking black coloration, while older specimens appear grey. The black widow tetra tolerates one of the widest ranges of pH of any aquarium fish, so should suit almost any setup. Keep in groups of six. Maximum adult size: 5cm (2in).

Harlequin Rasbora

Rasbora heteromorpha

The harlequin is a small lively fish from Southeast Asia that is ideal for all sizes of community aquarium. It has a distinctive black triangle on the rear of the body and shows its best colours when kept in a group of at least six. The fish prefer to swim in the upper half of the aquarium and enjoy the cover provided by floating plants. For a small fish, the harlequin has a huge appetite and will readily feed on flake food. Maximum adult size: 4cm (1.6in).

Pearl Danio

Brachydanio albolineatus

This is another peaceful species of danio and is ideal for a community aquarium. It is similar in shape to the zebra and leopard danio, with a pearly iridescence shot through with blue and purple. The females are larger and more colourful than males. Pearl danios originate from Southeast Asia and a pH of 6-7.5 would reflect their natural water conditions, but captive-bred specimens will tolerate a much wider range. They are best kept in a shoal but make sure your tank has a tight-fitting cover, as pearl danios like to jump. Colour varieties are available. Maximum adult size: females 6cm (2.4in); males slightly smaller.

White Cloud Mountain Minnow

Tanichthys albonubes

These agile little fish make a peaceful contribution to any aquarium. Their red-and-white fins look stunning against the darker body, which has a white stripe running along the flanks. Originally from China, they are some of the most commonly kept first community fish. These fish feel happiest in schools, as lone fish often become timid and lose their colour. They will tolerate a wide range of water conditions. Long-finned, veiled varieties have been bred, but often do not look as good as the original variety. Maximum adult size: 4cm (1.6in).

Fire-banded Barb
Barbus pentazona pentazona

The pentazona barb sports vivid stripes similar to those of the less sociable tiger barb. It constantly roams the aquarium looking for food and always seems interested in its surroundings. Beyond requiring good water quality during the maturation process, this very undemanding species is ideal for the beginner. It will thrive best in a group of five or six and is an excellent choice for all but the smallest community aquariums. Maximum adult size: 5cm (2in).

Cherry Barb
Barbus titteya

The cherry barb is one of the smallest barbs, but makes up for its lack of size with a stunning red body that serves it well in its natural habitat of shady Sri Lankan streams. These fish enjoy the cover provided by dense planting and the shade of floating plants. Introduce a group of five or six because cherry barbs can be timid if kept on their own. These peaceful fish are suitable for any community aquarium. Make sure their diet includes some flake food containing spirulina algae to keep the body colour bright. Maximum adult size: 5cm (2in).

The male's red coloration deepens during the breeding season.

Females are more orange in colour and have a deeper belly.

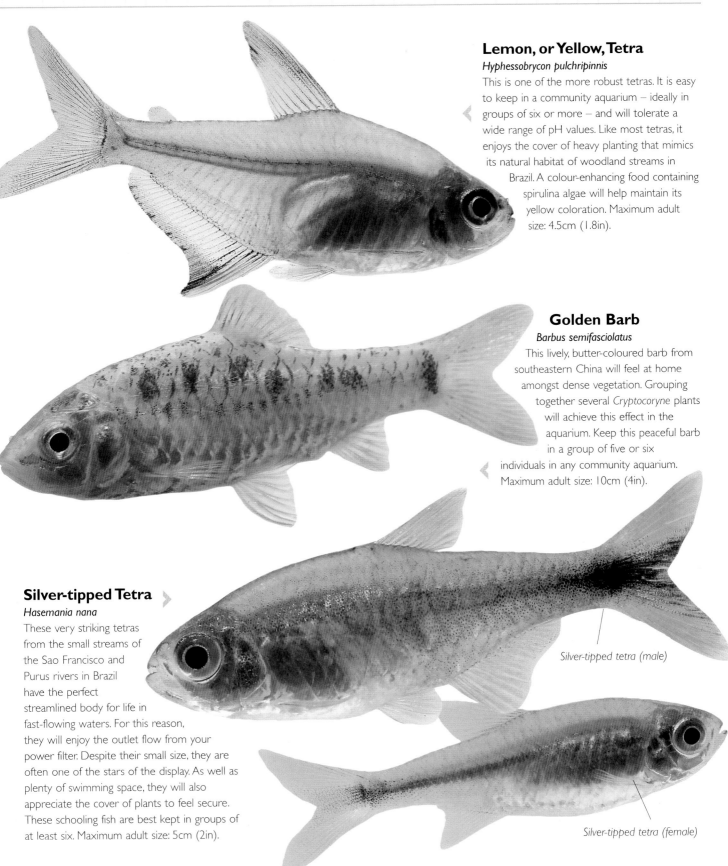

Lemon, or Yellow, Tetra
Hyphessobrycon pulchripinnis
This is one of the more robust tetras. It is easy to keep in a community aquarium – ideally in groups of six or more – and will tolerate a wide range of pH values. Like most tetras, it enjoys the cover of heavy planting that mimics its natural habitat of woodland streams in Brazil. A colour-enhancing food containing spirulina algae will help maintain its yellow coloration. Maximum adult size: 4.5cm (1.8in).

Golden Barb
Barbus semifasciolatus
This lively, butter-coloured barb from southeastern China will feel at home amongst dense vegetation. Grouping together several *Cryptocoryne* plants will achieve this effect in the aquarium. Keep this peaceful barb in a group of five or six individuals in any community aquarium. Maximum adult size: 10cm (4in).

Silver-tipped Tetra
Hasemania nana
These very striking tetras from the small streams of the Sao Francisco and Purus rivers in Brazil have the perfect streamlined body for life in fast-flowing waters. For this reason, they will enjoy the outlet flow from your power filter. Despite their small size, they are often one of the stars of the display. As well as plenty of swimming space, they will also appreciate the cover of plants to feel secure. These schooling fish are best kept in groups of at least six. Maximum adult size: 5cm (2in).

Silver-tipped tetra (male)

Silver-tipped tetra (female)

115

INCREASING THE FISH POPULATION

Once the aquarium has settled down and has been running with the first fish in it for a couple of weeks, you can think about adding more fish. As before, it is essential that you introduce any new fish a few at a time. This gives the bacteria in the filtration system a chance to increase gradually, so that they are able to break down the waste produced by the new fish. In addition to your usual testing regime, you must test for ammonia and nitrite levels regularly during the week following any fish introductions. This will allow you to take any remedial action necessary should the water quality deteriorate to an unacceptable level during this maturation process.

Another point to consider is the health of the new fish you are buying and whether you need to quarantine them at home in a separate tank. This was not necessary with the first fish, as they were 'quarantined' in the display aquarium. Even if they had been carrying any disease organisms, there were no fish there to pass them on to. The second – and any subsequent fish additions – are different, as they may bring diseases into the display tank and not only succumb themselves, but also pass on the problem to the existing healthy fish.

Five weeks of growth on the plants in your display will have started to show, however you should check the plants for any dead leaves or others that have started to yellow, as these should be removed to prevent them rotting in the aquarium. The plants will respond to the continued injection of carbon dioxide and the introduction of liquid feeds to continue to flourish and grow towards their full potential.

Right: *When transferring new fish from a quarantine tank to the main display aquarium, you may need to use a plastic bag rather than a net if the two systems are in different rooms. Use a clean bag and tape the corners as before.*

The existing fish are curious about their new companions, as these swim away from the plastic bag used to bring them from quarantine.

QUARANTINE

Even though you should be buying your fish from a reputable dealer who keeps them in the best possible conditions, it is a good strategy to quarantine any second and subsequent fish that you introduce.

The quarantine tank

Depending on the size of the fish you intend buying, a quarantine tank should measure at least 45x25x25cm (18x10x10in). Do not quarantine more than six 5cm (2in) fish at any one time in a tank this size, as you will not be able to control the water quality and maintain the excellent conditions required in the quarantine tank. If any individual fish is more than 10cm (4in) long, you will need a larger tank.

Like the display aquarium, the quarantine tank must be equipped with heating, filtration and lighting equipment, plus some aeration. An air-powered filter is ideal in a small aquarium, as it filters, aerates and circulates the water. It is also worthwhile including some decor to provide cover for the fish; very few species are happy in a bare aquarium.

Caring for the fish

When you bring the new fish home from the shop, add them to the quarantine tank, following the same procedure as described on pages 108-109 for adding the first fish to the display aquarium. From now on you must care for them exactly as if they were in the main aquarium, with daily feeding, weekly water changes and other regular maintenance to ensure that they remain in the best possible conditions.

Strictly speaking, you should keep a fish in isolation for longer than the longest 'incubation' period of any disease organism that could affect it. In real terms this could amount to a considerable period, as some viral diseases can lie dormant for a long time. For practical purposes, therefore, quarantine is dictated by the development period of the common pathogens most likely to affect aquarium fish. Generally, 20 days is an ideal length of time to isolate new fish. During this time they can recover from their stressful capture and journey home, and can be treated for any common pathogens if they show symptoms (see pages 192-193).

Monitoring the water quality

While the fish are in quarantine, check the water quality regularly. The pH level should match that of the main aquarium, unless a new fish has come from a retailer whose water is not the same as yours. In this case, alter the pH level until it is the same as in your display aquarium over the 20-day period. Do not subject fish to a daily change in pH of more than 0.3 of a unit. Maintain ammonia and nitrite levels at zero, with nitrate below 25ppm.

If there are no signs of disease after 20 days, you can transfer the fish to the main display aquarium. Always monitor the water quality in the main aquarium once the new fish have been added. To help the

A quarantine tank

Include a clean clay flowerpot or artificial decor item that will provide a refuge for nervous or poorly fish.

This simple internal filter is ideal for a quarantine tank. The twin sponges will sustain a useful population of beneficial bacteria to keep the water clean.

This airpump 'powers' the internal sponge filter

Use a heater-thermostat to maintain the water temperature at the same level as the main display aquarium.

A shallow bed of clean gravel will help the fish feel at home in their temporary quarters.

filter cope with the extra waste, add your filter start-up product a couple of days before you add the new fish. Follow the dosage rate recommended.

Maintaining the filter

Naturally, there will be times when the quarantine tank is not in use because you have not bought any new fish. However, the filtration bacteria still require a food source and one way of supplying this is to drop a small amount of fish food into the aquarium every two days or so. Another source of nitrogenous waste is some of the water in which you have cleaned the filter foam of the main display aquarium during routine maintenance. This will be rich in both waste and top-up bacteria for the quarantine tank filter. Before you quarantine more fish in the tank, be sure to check the water quality; aim for zero ammonia and nitrite readings.

Keep equipment separate

Keep a separate supply of equipment for the quarantine tank and never transfer any items used in the quarantine tank to the main display aquarium. You can quite easily pass a pathogen from sick fish to your main display aquarium on a wet net. If you do need to use any equipment in both aquariums be sure to disinfect it between each use.

Other uses

With all your fish healthily installed in the display aquarium, you might think that you no longer need the quarantine tank. However, there are several occasions when you will find it very useful.

If a fish becomes sick or distressed, it will benefit from the isolation of the quarantine tank, which then becomes a hospital tank. You will be able to inspect the fish closely and treat it as required before returning it to the main aquarium.

Some fish can become territorial, usually when breeding. They may pick on one of the more docile members of the display if it ventures too close to their eggs or offspring. An injured fish quickly

succumbs to secondary bacterial and fungal infections, so remove it to the sanctuary of the quarantine aquarium until the aggressive tendencies of the breeding pair subside. A bullied or injured fish can thus regain its strength before being returned to the main display.

If any fry are produced in the main aquarium, they will quickly come under attack from all the other fish (including even their parents), which regard them as a source of live food. If you wish to save the fry and grow them on, remove them to the quarantine aquarium. This can act as a nursery until they are large enough to be housed with other fish.

Fish swaps

Some retailers may be willing to take young fish off your hands and offer you a token credit for them in the form of aquarium equipment or food. Do not expect to get a retail price for them; after all, the retailer is doing you a favour and helping to find new homes for the fish. Always arrange a fish swap in advance to avoid moving (and therefore stressing) the fish unnecessarily if the retailer does not have space for them when you arrive at the shop.

Left: *Among the second fish to be added to our featured aquarium are some corydoras catfishes. As before, float the bag in the quarantine tank for 20 minutes to equalise the water temperatures before releasing them.*

Right: *Tilt the bag gently and allow the catfishes to swim away. A period in the quarantine tank will enable you to check the fish for any latent disease or health problem before they go into the main aquarium.*

Left: *One of the transferred catfishes explores its new home. Corydoras are popular bottom-dwelling fish that will use their sensitive barbels to locate food on and in the substrate.*

BARBS

The colourful and active barbs have long been popular choices for the home aquarium, but you should choose them with care to ensure their compatibility with the other fish in the tank. Retailers stock many species, and there is a wide range of body shapes, colours and temperaments to choose from. The cherry barb, for example, adds a flush of colour to the display and being one of the smaller barbs, is suitable for almost any aquarium. The tiger barb, on the other hand, while enjoying huge popularity because of its stunning colourful markings, will only be a suitable addition if you choose its tankmates with care. Nevertheless, in the right company it will become a focal point in the aquarium. Barbs readily shoal with members of their own species, and this behaviour adds considerable interest to any display. Be sure to keep shoaling species in groups of six individuals or more so that they feel secure. These fish will readily take flake foods. When the larger fish reach maturity, offer them pellets. You can offer freeze-dried live foods, such as bloodworm and brineshrimp, as a treat.

Arulius Barb

Barbus arulius

The arulius barb from India is a sociable species that prefers to be kept in a small group. For this reason, and also because it is one of the bigger barb species, it is best suited to the larger community tank. The arulius barb is placid, however, and can be kept with all but the smallest of tankmates. Males have a distinctive threadlike dorsal fin that adds a different texture to the display. As the fish matures, the body develops a purple-blue colour along the back, with black bars that act as part of the fish's natural camouflage. If the fish is not happy in the aquarium, the purple-blue fades to a dull grey. Maximum adult size: 12cm (4.7in).

Silver Shark (Bala Shark)

Balantiocheilus melanopterus

This stunning-looking fish from Thailand and Borneo gets its common name from its sharklike body shape. However, its temperament is peaceful and the silver shark can be kept with the smallest of fish. The distinctive silver body with black-edged fins makes a marked contrast to more colourful tankmates. This member of the carp family grows quite large, so is really only suited to tanks over 120cm (48in) long. Do not be fooled by small specimens offered for sale; they will grow and need a large tank with plenty of swimming space to be happy. Silver sharks are active and can be prone to jumping; be sure to provide a tight-fitting hood on the aquarium. Offer pellets to larger fish. Maximum adult size: 35cm (14in).

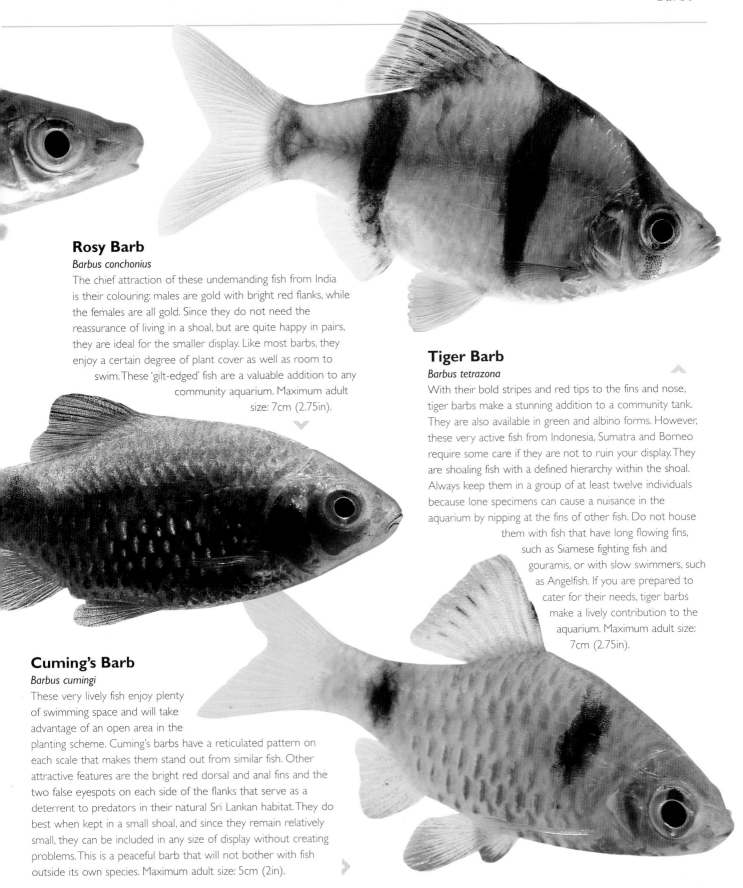

Rosy Barb
Barbus conchonius

The chief attraction of these undemanding fish from India is their colouring: males are gold with bright red flanks, while the females are all gold. Since they do not need the reassurance of living in a shoal, but are quite happy in pairs, they are ideal for the smaller display. Like most barbs, they enjoy a certain degree of plant cover as well as room to swim. These 'gilt-edged' fish are a valuable addition to any community aquarium. Maximum adult size: 7cm (2.75in).

Tiger Barb
Barbus tetrazona

With their bold stripes and red tips to the fins and nose, tiger barbs make a stunning addition to a community tank. They are also available in green and albino forms. However, these very active fish from Indonesia, Sumatra and Borneo require some care if they are not to ruin your display. They are shoaling fish with a defined hierarchy within the shoal. Always keep them in a group of at least twelve individuals because lone specimens can cause a nuisance in the aquarium by nipping at the fins of other fish. Do not house them with fish that have long flowing fins, such as Siamese fighting fish and gouramis, or with slow swimmers, such as Angelfish. If you are prepared to cater for their needs, tiger barbs make a lively contribution to the aquarium. Maximum adult size: 7cm (2.75in).

Cuming's Barb
Barbus cumingi

These very lively fish enjoy plenty of swimming space and will take advantage of an open area in the planting scheme. Cuming's barbs have a reticulated pattern on each scale that makes them stand out from similar fish. Other attractive features are the bright red dorsal and anal fins and the two false eyespots on each side of the flanks that serve as a deterrent to predators in their natural Sri Lankan habitat. They do best when kept in a small shoal, and since they remain relatively small, they can be included in any size of display without creating problems. This is a peaceful barb that will not bother with fish outside its own species. Maximum adult size: 5cm (2in).

Black Ruby Barb

Barbus nigrofasciatus

If tiger barbs appeal to you but you do not feel you can supply their needs, then black ruby barbs from Sri Lanka would make a wonderful alternative addition to your community aquarium. They have a similar body shape, with black vertical bars along the body. However, instead of red fins they have a rosy red glow that begins at the nose and spreads along each flank. The black fins are a contrast to the rouge body. The black ruby barb enjoys the cover provided by plant growth and can be timid if unable to find shade away from bright lighting. Prefers to be kept in a small shoal. Maximum adult size: 6.5cm (2.5in).

Clown Barb

Barbus everetti

The body shape of this attractive barb is slightly different to that of other community barbs in that it is more elongated. The fish is gold, with dark spots or bands along the flanks, while the red fins further increase the clown barb's appeal. Although it prefers the slightly acid water conditions found in its native habitat in Singapore and Borneo, your retailer may have specimens adapted to the local water conditions. Although non-aggressive, this lively barb will roam the aquarium, adding variety and interest. Prefers to be kept in a small shoal, but can be kept in pairs. Maximum adult size: 10cm (4in).

Checkered Barb
Barbus oligolepis

This wonderful little barb has a striking reticulated pattern on each scale that gives the flanks a checkerboard appearance. The males are a little bolder in colour, with reddish brown fins outlined in black. The fish look their best in a small shoal, patrolling the aquarium but never venturing too far from plant cover. The checkered barb from Indonesia and Sumatra is not only handsome, but also one of the most peaceful barbs, so well worth considering for any community display. Maximum adult size: up to 15cm (6in), but more commonly 5-7cm (2-2.75in).

Odessa Barb
Barbus ticto

Male Odessa barbs develop into stunning fish, with a bright red band along each side and a false black eyespot on each flank. Each scale has a reticulated pattern, so the flanks of the fish stand out well in the aquarium. Juvenile specimens offered for sale seldom display any adult coloration, so you must buy them in good faith, in the expectation that they will mature into handsome aquarium specimens. Although they will live happily in pairs, a small group of these peaceful barbs, which originate from India and Sri Lanka, will always make more of an impact in a community display. Maximum adult size: 10cm (4in).

Blue-Barred Barb
Barbus barilioides
Although this very lively fish from
Angola and Zimbabwe enjoys
swimming freely, it will spend a lot
of time under cover if the
aquarium lighting is too bright.
Given its placid nature, it should be
kept in a small shoal of eight or
more, otherwise individuals
become shy and will not thrive. The
vivid dark blue stripes are a
distinctive feature. This fish is ideally
suited to smaller community
aquariums, but it will need
generous swimming space.
Maximum adult size: 5cm (2in).

Black-Spot Barb
Barbus filamentosus
The black-spot is a beautiful small barb, very active
throughout the aquarium and worth considering as
an alternative to the tiger barb if you are worried
about the latter's potentially disruptive behaviour.
Juveniles have distinct two-bar markings on a gold
background, as well as bright red flashes on the
dorsal fin and tail lobes. As they grow, the fish lose
the black bar through the middle of the body, but
retain the black spot near the tail. Always look for
examples with two black bars; being the younger fish,
you can then enjoy watching their colours change as
they mature. These peaceful barbs from India and Sri
Lanka prefer the water to be slightly acid, but your
aquarium shop should have specimens acclimatised
to your local water conditions. Creates a better
display in a group of six or more. Offer floating
foods. Maximum adult size: 15cm (6in).

RASBORAS

Rasboras, a hardy group of fish from
Asia, are part of the carp family. Their
peaceful but active behaviour, along
with colourful markings, make them
ideal for the community aquarium. The
most common member of the family,
the harlequin rasbora, is an excellent
starter fish (see page 113). Other
members of the family are less
commonly seen in aquarium shops
but just as suited to a display. They
include the clown rasbora (*Rasbora
kalochroma*) and the dwarf rasbora
(*Rasbora maculata*). Rasboras are easy
to feed on flake food; they also enjoy
a treat of freeze-dried live foods, such
as bloodworm and brine shrimp.

Scissortail Rasbora
Rasbora trilineata

A unique tail pattern, combined with its swimming action, make this active fish look like a small pair of scissors swimming through the aquarium. The impression is reinforced when repeated throughout a small shoal and will make a strong impact in your display. The silver, black and white coloration of the scissortail contrasts well with the brighter colours of many other community fish. These rasboras from Malaysia, Sumatra and Borneo will enjoy open swimming spaces in your planting scheme. Do not mix them with boisterous barbs. Maximum adult size: up to 15cm (6in), but rarely more than 7-8cm (2.75-3.2in).

Clown Rasbora
Rasbora kalochroma

This sleek little fish, with its distinctive black spot marking, adds finesse to the display. However, despite their undoubted presence, the lovely clown rasboras can become a little possessive of a particular area within the aquarium. The solution is to keep them in a group of six or more with plenty of other fish, so that they do not get a chance to set up a territory. Their natural home is in Malaysia, Sumatra and Borneo and they prefer well-planted surroundings in which they can shelter from bright lighting. Maximum adult size: up to 10cm (4in), but rarely more than 7-8cm (2.75-3.2in).

Red-Striped Rasbora
Rasbora pauciperforata

These bright, lively individuals are best kept with fish of a similar disposition, as they feel more secure in numbers. However, you should also provide good planting so that they can reach cover should they feel the need. The bright red stripe along each flank adds to the pattern mix that other fish in the aquarium might not provide, so this rasbora is worth adding to any community display – ideally in a group of eight or more. Originating from Southeast Asia, Malaysia and Sumatra, it prefers slightly acid water, but your local retailer may have some examples that have been acclimatised to your local water conditions. Offer softened algae wafers in addition to the usual aquarium foods. Maximum adult size: 7cm (2.75in).

DANIOS

Danios are among the hardiest aquarium fish and an excellent choice for newcomers to the hobby. They are ideal starter fish, easy to feed and look after. Danios are always busying themselves around the tank in search of food and this lively behaviour makes them very endearing subjects for community aquariums. They will not eat any of the aquarium plants and will not grow large enough to disrupt any aquarium decor. There are several species to choose from, but the zebra and the leopard danio (page 112) are the most frequently sold. Long-finned varieties of both have been bred, but are not as appealing as the natural forms of the danio or other species of fish with naturally long fins. Danios are best kept in a small shoal to help them feel secure.

Giant Danio
Danio aequipinnatus

As its name suggests, this is a larger species of danio and much deeper-bodied than its slender cousins, the Zebra and Leopard Danio. It sports lovely trout-style markings down each flank on a background of turquoise-blue that stands out in the aquarium. The Giant Danio, from Malaysia, Sumatra and Borneo, has an upturned mouth, an indication that it often feeds from the surface in the wild. In the aquarium, it will readily adapt to manufactured floating foods. This lively, active species is happiest living in a small shoal in an aquarium at least 1m (39in) long. Make sure it is securely covered, as these active fish are liable to jump out. Maximum adult size: 10cm (4in).

The giant danio is a fast-swimming surface dweller.

Gold giant danio, a colour form bred for the aquarium hobby.

TETRAS

Most of the following species are naturally shoaling. so to create the best effect, include a dozen or more individuals in a single display. Do not add them all at once, however, as the water quality will suffer. To avoid major problems, introduce them in small groups of three or four individuals, thus allowing the filter to adjust gradually to the extra load of new fish.

Emperor Tetra
Nematobrycon palmeri

For something really different, consider adding the stunning emperor tetra from Colombia to your aquarium. It is one of the few fish that will add a touch of purple to the display, a colour that stands out against the greens and reds of the plants. Its aggressive-looking expression hides a placid nature, so do not keep it with any large, boisterous fish. For a tetra, the emperor is a long-lived fish, surviving up to six years in a well-cared-for display. Maximum adult size: 5cm (2in).

Cardinal Tetra
Paracheirodon axelrodi

In the cardinal tetra, the red band runs the entire length of the fish, making it slightly more colourful than the neon. For a special display in a suitable aquarium, keep these fish in a large shoal of 25 to 30 or more. They look best with little in the way of other midwater swimmers, giving the cardinals exclusive rights to this area of the aquarium. Then the fish will exhibit the natural shoaling behaviour seen in their native Brazilian habitat. They can be included in a mixed aquarium (but not with larger, aggressive fish), where a minimum of six individuals will add to the overall display, yet still feel secure. They are happier with softer, more acidic water. Maximum adult size: 5cm (2in).

Neon Tetra
Paracheirodon innesi

This stunning little tetra from Peru brings visual delight to a display and is probably the most popular aquarium fish ever kept. The almost unnaturally vibrant red and neon-blue flashes are seen to best effect when a group of at least six individuals demonstrate their natural shoaling behaviour, and a brilliant display can be achieved with a dozen or more. Do not house them with larger fish that may pick off the neon tetra as a dietary supplement. Angelfish, for example, have a habit of doing this, so do not keep them in the same aquarium. Maximum adult size: 4cm (1.6in).

Bleeding Heart Tetra
Hyphessobrycon erythrostigma

The red spot just below the lateral line gives the bleeding heart tetra its common name, and the unusual diamond-shaped body adds another dimension to the aquarium display. These large tetras from Peru can be kept in a pair instead of a group, but in a large tank they certainly make an impact in a shoal. Males can be identified by extended dorsal and anal fins. The fish prefer water with a pH level between 6.5 and 7.2. However your aquarium shop may have specimens acclimatised to your local water conditions. Bleeding heart tetras prefer a quiet aquarium, so avoid including them with busier, larger fish and potentially aggressive species. Maximum adult size: 6cm (2.4in)

Glowlight Tetra
Hemigrammus erythrozonus

The glowlight tetra from Guyana adds a brilliant flash of orange-pink to the aquarium display. The unusual coloration (for a tetra) makes this placid fish stand out from the crowd and explains why it is so popular. The colour stripe runs the entire length of the body, including parts of the eye and tail. The colours look their brightest against the dense planting these fish prefer. Keep the glowlight in a shoal of at least six individuals. Maximum adult size: 4cm (1.6in).

Head-and-Tail Light Tetra
Hemigrammus ocellifer

The common name of this peaceful tetra from Guyana and Bolivia comes from the two shiny patches of skin close to the head and tail. As with most tetras, the head-and-tail light prefers to live in a shoal of six to eight individuals. In ideal conditions, these fish are likely to spawn in the community aquarium. Maximum adult size: 4.5cm (1.8in).

Black Neon Tetra

Hyphessobrycon herbertaxelrodi

Although distinctly different from the neon tetra, the black neon from Brazil is similar in body shape. It sports a pale green stripe alongside a predominantly black one and these unusual colours are complemented by red on the top half of the eye. The black neon is more demanding in terms of aquarium conditions than most of its cousins. Ideally, the pH should be on the low side for a community aquarium – between 6.5 and 7.5 – but your aquarium shop may stock specimens that are acclimatised to your local water conditions. Keep a minimum of six individuals. Maximum adult size: 4cm (1.6in)

Serpae Tetra

Hyphessobrycon callistus

The serpae tetra is a bright-red, active fish that brings constant movement to the display. Being undemanding in terms of conditions, it may represent a better option than the red phantom tetra. However, being more boisterous, it should be kept in a shoal of at least six or more individuals to ensure that any bickering remains within the shoal and does not affect other fish in the display. This tetra from Paraguay makes a good, colourful addition to any aquarium. Maximum adult size: 4cm (1.6in).

Flame Tetra
Hyphessobrycon flammeus

The male flame tetra is particularly striking, with a bright red anal fin and lower body, while the rest of the body is a dusky red. The females are more subdued in coloration. This easy-care, peaceful fish is worth considering as an alternative to the red phantom tetra, which can be more delicate. Do not be deterred if these peaceful Brazilian fish do not show their full colours in the shop; once settled in a planted aquarium, they will colour up. Keep a minimum of six individuals. Maximum adult size: 4cm (1.6in).

Red-Eyed Tetra
Moenkhausia sanctaefilomenae

The upper half of this fish's eye is bright red, hence the common name. But this bold-looking tetra also has a distinctive black band across the tail and dark-edged scales that give it an almost armour-plated appearance. These lovely shoaling fish from Brazil, Bolivia and Peru will swim towards the surface as well as in midwater and enjoy the cover provided in a well-planted aquarium. Being slightly bigger than most tetras, they are best suited to a larger aquarium if you plan to keep a shoal. Maximum adult size: 7cm (2.75in).

Congo Tetra
Phenacogrammus interruptus

The Congo tetra from Zaire is a stylish addition to the aquarium. Its gold and petrol-blue flanks are complemented, in the males only (shown here), by long flowing fins. It is one of the largest popular tetras, but despite its size, it is timid and should not be kept with boisterous fish. Congo tetras are best kept in a small shoal of six individuals in an aquarium measuring at least 90cm (36in), as they need plenty of swimming space. Provide an environment with relatively dense planting, although not fine-leaved plants, which they will nibble. Some surface plant cover will make the Congo tetras feel at home and encourage them to show off their colours. Maximum adult size: Males 8.5cm (3.3in), females slightly smaller.

Buenos Aires Tetra
Hemigrammus caudovittatus

This bold, colourful tetra from Argentina, Paraguay and Brazil is easy to keep and longlived by tetra standards, making it an excellent choice for the first-time aquarium keeper. All the fins, other than the dorsal, are bright red and the body is silver. A black bar runs from the base of the tail and between the lobes. The Buenos Aires tetra provides constant colourful movement in the aquarium. Its only potential drawback is a tendency to eat fine-leaved plants, so only keep it in a display with robust plants that can withstand some attention. Keep a minimum of six individuals. Maximum adult size: 7cm (2.75in).

GOURAMIS

Gouramis are a peaceful group of fish that swim in the middle and upper layers of the aquarium. As most species have long, flowing fins, they should not be kept with other fish that have a tendency to fin nip. In the wild, gouramis live in swampy wetlands, where dissolved oxygen levels are low. In addition to using their gills, they also take in air at the surface and pass it into a convoluted labyrinth organ near the gill chamber that allows oxygen to pass into a network of tiny blood vessels. Gouramis share this ability with other 'labyrinth fish' such as Siamese fighting fish.

Chocolate Gourami

Sphaerichthys osphromenoides osphromenoides
With its delicate cream stripes set against a chocolate-brown body, this shy fish adds an element of finesse to the display. In the wild, its coloration provides excellent camouflage, so for the fish to feel happy in the aquarium, it needs the cover provided by lush planting. Chocolate gouramis also benefit from being kept in a minimum of a pair, and given their timid nature, they should only be kept with other peaceful species. They can be difficult to keep. They prefer the slightly acid and soft waters found in their native habitats in Malaysia, Sumatra and Borneo, but your retailer may have some specimens that have been acclimatised to local water conditions. Maximum adult size: 5cm (2in).

Pearl Gourami

Trichogaster leeri

The stunning coloration of the pearl gourami from Malaysia, Sumatra and Borneo helps to explain why it is one of the most popular gourami species. Each spotted flank has a jet black line running from the nose through the eye and all the way to the tail. As they mature, males (shown here) develop a red breast, the colour stretching from the underside of the mouth to the start of the anal fin. Males also have extended, almost pointed dorsal fins, while those of females remain rounded. This will help you to select a pair, although most retailers only sell them as a pair. Pearl gouramis can be quite long-lived for aquarium species, surviving up to eight years, but they must be kept with other peaceful species if they are to fulfil their potential. They prefer to occupy the upper layers of the tank and enjoy the shelter of reasonably dense planting. Maximum adult size: 12cm (4.7in).

Thick-Lipped Gourami

Colisa labiosa

The thick lips that give rise to this fish's common name are ideal for sucking in floating food and, indeed, it happily occupies the upper and middle parts of the aquarium. Its coloration is very similar to that of the dwarf gourami, with diagonal stripes of red and cobalt blue, but the thick lips are a clear distinguishing feature. Males (shown here) have the pointed dorsal fin seen in most gouramis, making the fish relatively easy to sex. Thick-lipped gouramis, native to India and Myanmar (Burma), are among the larger community species of gourami available. Provide slightly acid water conditions and keep with peaceful species only. Maximum adult size: 9cm (3.5in).

Dwarf Gourami

Colisa lalia

The dwarf gourami, from India and Borneo, is one of the all-time classics for the small to medium-sized community aquarium. The colour variants are immense, but the most popular are those with the natural diagonal red-and-cobalt blue stripes. Like other gouramis, they appreciate the protection of dense planting, especially the extra cover created by floating plants. They dislike being hassled by boisterous tankmates, so choose your stock carefully. Dwarf gouramis thrive better when kept in pairs. Females tend to have less vibrant colours than males (shown here), so ask your retailer to select a pair. Maximum adult size: 5cm (2in).

Moonlight Gourami

Trichogaster microlepis

This 'ghost' of the aquarium is a pure silvery white, and like all gouramis, it has very elongated pelvic fins. In good specimens, these will carry the only other colour seen on these fish: orange to red in males, yellow in the females. However, a much easier way to sex a pair is to identify the male by his extended, pointed dorsal fin. The moonlight gourami can become the target of fin-nipping fish, so keep it with peaceful species only and provide this timid fish from Thailand and Cambodia with the cover provided by dense planting. Maximum adult size: 15cm (6in).

Snakeskin Gourami

Trichogaster pectoralis

Small scales cover the body of *T. pectoralis*, hence its common name. The principal marking is the black stripe along the lateral line on each side of the fish. This is broken by some vertical silver stripes near the tail. In males, there is a red hue to the anal fin, while in females (shown here), it is yellow. The easiest way to identify a pair, however, is to look for the pointed dorsal fin of the male. The Snakeskin, from Malaysia, Thailand and Cambodia, grows relatively large, so is ideally suited to the larger, peaceful community aquarium. It is one of the most placid species. Maximum adult size: up to 20cm (8in).

WARNING: One gourami to avoid

If you are ever tempted to buy a fish labelled as the common gourami (*Osphronemus gorami*), then please think again! *Osphronemus gorami* is a real giant of the fish world, growing to 70cm (28in) or more. Although it is a beautiful fish in its own right, it should only be kept in public aquariums and by fishkeepers who are dedicated to providing the necessary space and investment that these fish require. When small, *Osphronemus gorami* could be mistaken for a large chocolate gourami, so choose with care and ask your retailer for positive identification.

Croaking Gourami

Trichopsis vittata

The slender-bodied croaking gourami brings another dimension to the community aquarium, since both the male and female make croaking noises when displaying. They also add a rich tapestry of colours, including the familiar reds and cobalt blues. These timid fish from India, Thailand, Vietnam, Malaysia and Indonesia need surface planting for cover. Maximum adult size: 6.5cm (2.5in).

Three-Spot or Blue Gourami

Trichogaster trichopterus

The three-spot gourami makes a distinctive addition to an aquarium display. Two of the spots are black body markings against the blue skin, while the third is the eye. Together, these form its delicate camouflage. It is one of the most commonly available aquarium species and also one of the most sedentary. The males can be quite aggressive, however, so choose fellow species with care. The three-spot gourami, from Malaysia, Burma and Vietnam, will tolerate most water conditions. Selective breeding has produced a golden form that does not have spots and is sold as the golden gourami. Maximum adult size: 10cm (4in).

Honey Gourami

Colisa sota

Being one of the smaller gourami species, the honey gourami from India, Assam and Bangladesh is ideally suited to the smaller setup. The glorious honey colour is seen in the male, which also has a dark blue-black head and belly. To thrive, the honey gourami, prefers dense plant cover in which to hide. Like all gouramis, these fish can become territorial when they breed, so the plant cover helps to protect both them and other fish in the aquarium. Keep with peaceful species only. Maximum adult size: 5cm (2in).

CATFISHES

Catfish are a fascinating group of predominantly bottom-feeding fish that add movement to the lower area of the aquarium, and their varied patterns of behaviour are a source of endless interest to the fishkeeper. For example, the busy *Corydoras* catfishes spend a great deal of their time picking over the substrate looking for food, while the suckermouth species often sit and watch the proceedings in the tank before shuffling around to another vantage point. For balance and interest, suitable catfish should be included in every display.

Corydoras

Even though nearly 200 *Corydoras* species are known to science, only a few dozen are available to aquarium hobbyists. The small size and placid nature of these South American catfish (many hobbyist fish are originally from Brazil) make them ideal community subjects, and their busy nature and often striking markings add further to their appeal. They are all peaceful species that can be mixed and matched as you wish, but in common with most fish, they create the most natural display and impact when seen as a shoal of the same species. Indeed, *Corydoras* never seem to thrive as individuals, so it is always a good idea to buy a small group. If you intend keeping *Corydoras* catfishes in your display, be sure to choose a smooth substrate material, as any sharp edges will wear away the barbels that they use to help them find food. Offer them tablet and wafer foods, with treats of freeze-dried livefoods, such as bloodworm and tubifex.

Corydoras duplicareus

This small *Corydoras* catfish from the Amazon River system can be difficult to find in aquarium shops but is worth pursuing. The body is a light sandy colour with a jet black stripe down the back and another across the head to hide the eye. In between the two there is a stunning gold spot that contrasts well with the black markings. This *Corydoras* will take most foods, but include treats of freeze-dried or frozen bloodworm. Ensure your display has a section of open substrate on which these fish can forage. This peaceful species is a great addition to any community display.
Maximum adult size: 5cm (2in).

Bronze Corydoras

Corydoras aeneus
Sporting deep bronze, armoured scales, the bronze corydoras is a common sight in aquarium shops and an excellent choice for the beginner. Its industrious behaviour is typical of this group; it enjoys scouring open areas of substrate for food and will keep foreground plants free of debris. This firm favourite with aquarium keepers worldwide is also available in an albino form, which makes a fine alternative to the bronze coloration.
Maximum adult size: 7cm (2.75in).

Artificially coloured fish

When choosing fish, do not accept any albinos that have been artificially coloured, as the colouring process drastically reduces their life expectancy. By rejecting these fish, the unethical process of fish colouring can be discouraged.

Harald Schultz's Corydoras

Corydoras haraldschultzi

This species closely resembles *C. sterbai*, although the colours are usually much darker. To tell them apart with certainty, look closely at the spots. *C. haraldschultzi* has dark spots on a light background, while *C. sterbai* has light spots on a dark background. In common with *C. sterbai*, it has yellow-gold front rays on the pectoral and pelvic fins, which can darken to orange in mature adults of both species. Maximum adult size: 7cm (2.75in).

Corydoras gossei

This beautiful little catfish has a mottled head pattern, while the tail has an arrangement of broken stripes. The dark flanks contrast with the very pale underside and the whole body is highlighted by bright yellow leading edges to the dorsal and pectoral fins. *C. gossei* prefers slightly acid water, but your local shop may have some examples that are acclimatised to your local water conditions. In the aquarium, it looks even better when kept together with a group of *C. sterbai*. Maximum adult size: 6cm (2.4in).

Peppered Corydoras

Corydoras paleatus

Along with the bronze corydoras, the peppered corydoras is another highly popular catfish and commonly available in the aquarium trade. The mottled patterns along each flank act as perfect camouflage in the wild. The white underside of the peppered corydoras is often visible as the fish scurry around the aquarium. An albino form is also available. Maximum adult size: 7cm (2.75in).

Corydoras reticulatus

The marbled pattern of this Peruvian corydoras works as effective camouflage in the wild, and adds an impressive element to the aquarium display. The dark flanks have silver markings, the dorsal fin is black with a silver stripe and the tail is also striped. *C. reticulatus* is one of the most attractive *Corydoras* species and suited to all but the smallest of displays. Keep some of the foreground substrate free of plants to allow the fish to forage and you can enjoy observing a small group bustling round the aquarium.
Maximum adult size: 7cm (2.75in).

Dwarf Corydoras

Corydoras pygmaeus

Most *Corydoras* species inhabit the lower reaches of the aquarium, but unusually, *C. pygmaeus* spends most of its time swimming in the mid to upper layers. The fish are best kept in a group of half a dozen and are easy to identify by the black stripe that runs from nose to tail. In most *Corydoras* species the mouth is underslung, but in *C. pygmaeus* it is more forward facing. When it comes to feeding this baby of the *Corydoras* group, it is a good idea to crush up some flake food, so that it can feed easily in midwater. Watch the fish carefully to ensure that they receive enough food amongst the other busy midwater swimmers.
Maximum adult size: 2.5cm (1in).

Panda Corydoras

Corydoras panda

The panda is one of the smaller species of *Corydoras*, but easy to recognise by the black bands on the head, dorsal fin and caudal peduncle. These markings, set against a pale background, produce a striking and well-named addition to the display. *C. panda*, from Peru, has been known to spawn in the aquarium, but picking males and females may be difficult in the aquarium shop. Mature females are usually larger than males, but this difference will not be apparent in young fish. The best option is to buy a small group in the hope of acquiring at least one of each sex. Maximum adult size: 4.5cm (1.8in).

Sterba's Corydoras

Corydoras sterbai

One of the stars of the *Corydoras* family, *C. sterbai* has a beautiful mottled pattern with white spots on its head and broken stripes along each flank that extend into the tail. This striking pattern is enhanced by the stunning golden leading edges to the pectoral and pelvic fins. To help the fish feel at home, make sure they have an area of open substrate to swim around. Maximum adult size: 8cm (3.2in).

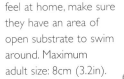

LOACHES

Although these fish belong to the same family as barbs and danios, they have evolved to live on the floor of their chosen habitat. The loaches forage in the aquarium substrate, feeding on particles of food missed by other fish. They still need feeding and will enjoy tablet and wafer foods, plus treats of freeze-dried or frozen bloodworms. They have underslung mouths surrounded by sensitive barbels that enable them to find their food by touch and taste. Loaches are generally peaceful, but some species should be avoided, as they will hassle other tankmates. The following species are ideal for a community display.

Clown Loach

Botia macracantha

A long-time favourite of aquarium keepers, the clown loach from Southeast Asia is a stunning bottom-dwelling fish. The bold black-on-gold striped flanks, complemented by scarlet fins, are only part of its attraction. It is also a very active fish, only shying away from the brightest lights in the aquarium, so provide some hiding places. It accepts the usual tablet and wafer foods, plus a treat of freeze-dried or frozen bloodworms, but will add to its diet by picking off unwanted snails, thereby helping to keep the plants in good condition. The clown loach can be long-lived and mature specimens can reach a large size for a community aquarium, although this will take several years. Maximum adult size: 16cm (6.2in).

Dwarf or Chain Loach

Botia sidthimunki

The dwarf loach, from India and Thailand, has a very pretty chain pattern along each flank that runs all the way from the characteristic underslung mouth to the tail. It is an ideal bottom-dwelling community fish, constantly active over the substrate looking for food and, like the clown loach, will perform a valuable service in picking off unwanted juvenile snails. Keeping these fish in a small group of four or five helps them to feel secure. For the same reason, you should make sure they have some hiding places. The dwarf loach's mature size makes it an excellent alternative to the clown loach if you are worried about the latter's potential size. Maximum adult size: 5.5cm (2.2in)

Guppies, Mollies, Platies & Swordtails

As their name suggests, livebearers produce live young, rather than lay eggs, which is the typical method of reproduction among fishes. These prolific fishes are often among the first fish to breed in an aquarium. They are a varied group of fishes, many of which are ideal candidates for a community tank. Their busy nature often takes them to all corners of the aquarium, which adds to their appeal. Livebearers are easy to keep and relatively undemanding in their requirements, making them ideal beginner fish. As with all livebearers, the male can be distinguished by its gonopodium, a tubelike modification of the anal fin used in mating.

The shimmering scales of the neon blue guppy sparkle in the aquarium lights.

Guppy
Poecilia reticulata

It is not hard to understand why the stunningly colourful and active guppy should be one of the all-time favourites of the community aquarium. The long flowing tails of the males have vibrant patterns in colours ranging from bright red to mottled green and yellow. Many guppy varieties have been bred to add to the naturally occurring colour patterns. Selective breeding has also provided the fishkeeper with a range of tail shapes and sizes to choose from. As virtually all specimens offered for sale are farmed fish, the guppy, originally from Central America, is adapted to most aquarium environments. However, in soft water areas you may need to raise the hardness to ensure that the pH level does not drop too low. The peaceful guppies occupy the mid and upper layers of the tank and are ideal for the community aquarium, but avoid keeping them with boisterous barbs and Siamese fighting fish that will nip their flowing tails, leaving them stressed and very susceptible to disease.
Maximum adult size: 6cm (2.4in).

The king cobra colour form combines a beautifully mottled body with a banded tail.

The golden form glows in shades of yellow and orange.

Black Molly
Poecilia sphenops

The black molly is commonly available in aquarium shops and, although not as flamboyant as its sailfin cousin, it makes a lively addition to the middle and upper water levels of any community display. Its placid nature means that it will co-exist happily with all other peaceful species. It prefers reasonably hard water, with a pH of over 7.5. Maximum adult size: 6cm (2.4in).

Male and female green sailfin mollies.

Sailfin Molly
Poecilia velifera

The dorsal fin of the male sailfin molly is emblazoned with a multicoloured mottled pattern and edged with gold. The pattern extends onto the tail and complements the scale pattern on each flank. The fish are usually sold in sexed pairs, and the male uses his magnificent dorsal fin to 'show off' to his female. Whilst the natural colour of this Mexican fish is a green-yellow-silver mix, coloured varieties are also available, including gold and silver. Mollies are active fish that patrol the middle to upper layers of the aquarium for food. The upturned mouth is characteristic of fish that feed from the surface in the wild, so floating flakes pose no problems in the aquarium. Despite its showy behaviour, the sailfin molly is peaceful towards other tankmates, but for its own sake, do not keep it with potential fin-nippers. Ideally, keep mollies in hard water with a pH above 7.5. Maximum adult size: 15cm (6in).

An elegant male orange sailfin molly.

Swordtail

Xiphophorus helleri

The bright and colourful Central American swordtails bring a unique body shape to the community aquarium. Although the sword is only found on the males, females are just as brightly coloured and usually larger than males. Swordtails are generally peaceful fish that appreciate the security of aquarium plants as they swim in the middle to upper water levels. If you are keeping more than one male, make sure there is plenty of cover, as they may become antagonistic towards each other. Swordtails are available in many colour varieties. These have proved easy to develop, given the fishes' readiness to breed, which they will do in your aquarium. Maximum adult size: males (excluding sword) 10cm (4in), females 12cm (4.7in).

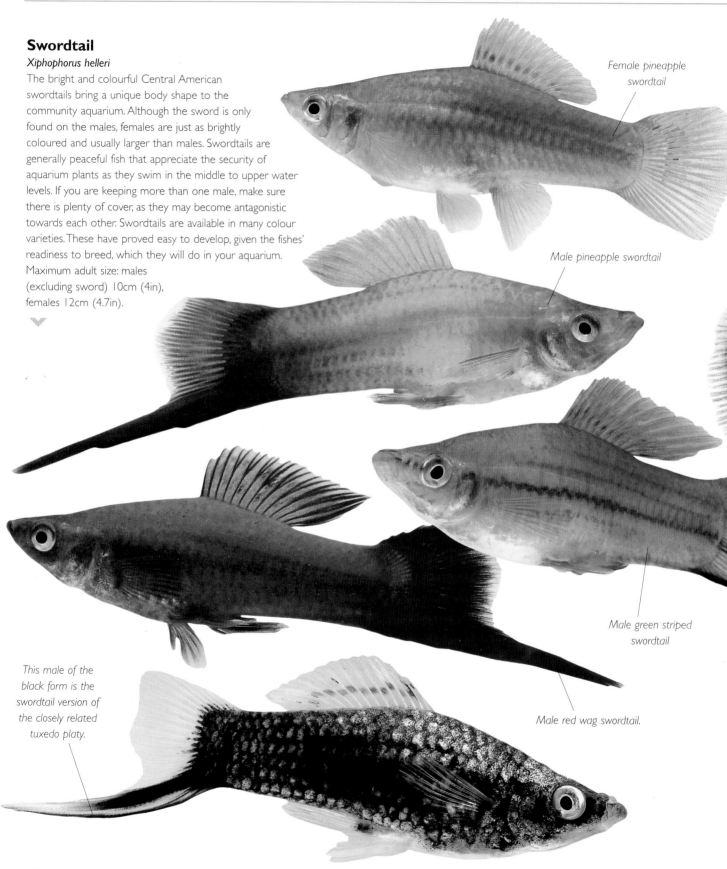

Female pineapple swordtail

Male pineapple swordtail

Male green striped swordtail

Male red wag swordtail.

This male of the black form is the swordtail version of the closely related tuxedo platy.

Platy
Xiphophorus maculatus

The sociable and colourful platy is ideally suited to the community aquarium, although long-finned varieties should not be kept with boisterous barbs or Siamese fighting fish. Bear in mind that platies may eat fine-leaved plants, although broadleaved, larger plants will be safe from their attentions. A worthy addition to any display, with almost endless colour varieties. Maximum adult size: males 3.5cm (1.4in), females 6cm (2.4in).

Male blue one-spot platy

Male orange platy

Female tuxedo platy

Male red wagtail platy

Right: *Two colour forms of platies swim together in this aquarium. The two males at the top are hifin types, with a larger dorsal fin than normal forms. The two females below are sunset forms.*

TOWARDS A STABLE SYSTEM

After seven weeks the featured aquarium is making good progress. The plants have grown to cover the equipment on the lefthand side, and in time, the filter on the right will also be hidden from view. The plants are also filling the spaces left for them when they were first put in, and individuals of the same species, such as the hemianthus and cryptocoryne, now look like single specimens. A real sign that the plants are happy is that during the day they will photosynthesise so rapidly that they produce streams of oxygen bubbles from their leaves.

The plants are framing an open area towards the centre of the aquarium, where we can view the fish as they play in the current created by the internal power filter. Their lively performance is proof that the effort made to provide them with a stable environment in these crucial early weeks has paid off; they are healthy, happy and thriving.

To keep the aquarium in good shape, it is time to consider some of the more long-term maintenance tasks that should become part of your regime. For example, the equipment you have installed will not run forever without some attention, and cleaning it must now be added to your list of regular tasks. Again, this is not difficult and does not take long, but is essential if the aquarium is to remain healthy and looking good.

If algae growth is proving to be a nuisance, it is vital to understand what stimulates it and how to eliminate it if possible. As water testing is the only way to gain a true insight into the aquatic habitat you are controlling, use the different tests available to explain any problems in the aquarium. For example, an excess of phosphate in the tapwater supply might account for an unexpected algae bloom; the only way to confirm your suspicions is by carrying out a water test.

It is now five weeks since the first fish were added and three weeks since the second fish went in. Most of the plants are thriving, but there is evidence of an algae problem. This is a critical time.

CARE AND UPKEEP

As we have seen during the previous weeks, it is important to adopt a regular maintenance routine if the display is to look its best and the environment is to remain healthy. Before embarking on any maintenance, assemble all the equipment you will need, including a clean bucket for the water changes, sharp scissors for trimming plants and a bin for debris. Turn off the electricity supply. Leave yourself enough time to carry out the work slowly and carefully in order not to disturb the plants and substrate too much.

Clean the glass on a weekly basis. As time passes, the plants growing near the edges of the tank will require extra attention, otherwise they will grow too large and restrict easy access to the glass for cleaning. You may also need to move the substrate near the front of the aquarium to clear away any algae growing on the inside of the front glass below the substrate level. Do this very carefully to avoid damaging any plant roots. If in doubt, you can leave the algae below the

Above: *There is evidence of algae on the front glass. Wipe this away with a wad of filter wool and clean the remainder of the glass to give you a clear view of what is going on inside the tank.*

Testing for phosphate

Phosphate is present in tapwater and is also gradually added to the aquarium via fish food. It can promote sudden unwanted algae growth and must be controlled to prevent the display from being ruined. After running for seven weeks, the featured tank has a high phosphate level of over 2.0ppm. This was suspected when the telltale signs of algae began to appear. Do not panic if this should happen; simply carry out water changes using water containing less phosphate to help reduce the overall level. A chemical resin filter media will rapidly reduce this level, starving the algae out of existence.

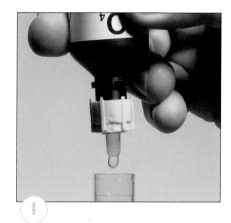

Take a sample of water from the aquarium and slowly add eight drops of reagent from the yellow bottle.

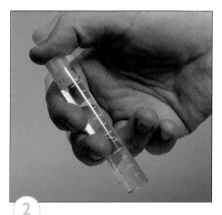

Put the stopper on the test tube and shake the tube thoroughly. Leave the solution to develop for the recommended time.

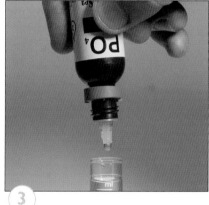

Add two drops of reagent from the green bottle to the solution. Store opened containers of reagent as instructed.

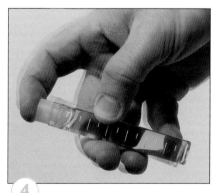

Shake the test tube again and allow it to stand as directed. Always follow the manufacturer's instructions exactly.

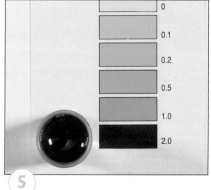

Remove the stopper and, looking down the tube from above, compare the colour of the solution against the chart provided.

Algae spreads through the tank

High phosphate and nitrate levels plus excess light can boost algae growth. To solve the problem in this tank, the rate of water changes was increased for two weeks from one 10% change per week to two 10% changes per week. The water going back into the tank had less phosphate and nitrate in it. The lighting was reduced by half an hour at the beginning of the day. The algae soon died back, leaving a dusting of brown sludge that was removed. If husbandry changes are not enough, there are specific treatments suitable for use with aquarium plants. The better ones will alter the water chemistry, removing essential salts and preventing algae from growing. These changes have little or no effect on the plants in the aquarium.

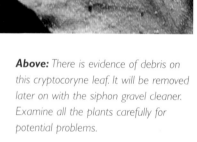

Above: There is evidence of debris on this cryptocoryne leaf. It will be removed later on with the siphon gravel cleaner. Examine all the plants carefully for potential problems.

Below: Algae on the aquarium glass is not unusual. Algae problems can arise seemingly overnight, even in well-managed aquariums. Taking prompt action should keep them under control.

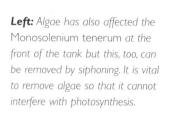

Left: Algae has also affected the Monosolenium tenerum at the front of the tank but this, too, can be removed by siphoning. It is vital to remove algae so that it cannot interfere with photosynthesis.

substrate level alone, otherwise you may do more harm than good.

Once you have had your hands in the water, you may need to wipe up drips on the outside of the aquarium. Only use aquarium glass cleaners, as they have been formulated to be safe with livestock. Most general household chemicals are toxic to fish and should not be used anywhere near the tank.

Trimming the plants

As we discussed in week three (see page 104), it is necessary to prune the display plants to keep them in check and stop them from outgrowing their allotted space. It is also necessary to prune and remove older growth that starts to die off, as this will very quickly rot in the aquarium adding to the organic waste load on the biological filtration system. This is the case in the featured tank, where several of the leaves on the *Echinodorus* have started to become skeletal as the old tissue dies and starts to decay. Be sure to trim these stems as near

Above: *If one plant is taking over at the expense of another, or just generally spreading beyond its allocated growing space, do not be afraid to trim it back, as here with hygrophila.*

Left: *Hemianthus will benefit from a regular 'haircut' to encourage new, bushy growth and keep it tidy. One thorough trim is better than snipping away piecemeal every few days.*

Above: *As you trim the hemianthus, the ends will float off around the aquarium. Remove them with a fine net when you have finished cutting.*

Below: One leaf of Echinodorus 'Red Flame' has completely died back and should be removed as soon as possible. It is unsightly and of no further use to the plant.

Right: This echinodorus leaf at the back of the aquarium is also beginning to show signs of damage. It is best to cut it out now to make room for new leaves.

Cleaning the CO2 pump

You will need to clean the small CO2 dosing pump. The small bits of leaf from trimming the hemianthus have been sucked into the inlet; remove these with your fingers. Every two months take the pump out of the aquarium, rinse out the impeller and check for wear.

Below: Plant trimmings may become lodged in the intake of the CO_2 water pump. Remove them with your fingers.

Below: Using sharp scissors, trim the leaf away cleanly at the base of the stalk and remove it from the aquarium. Plants will continue to look their best if you maintain them regularly.

Above: *In a water change, the aim is to replace 10-20% of the water. This photo shows the water level after the siphoning shown opposite.*

Right: *Using a siphon cleaner, you can make a water change and remove algae and debris from plants, such as* Monosolenium, *at the same time.*

as possible to the crown of the plant, leaving very little to rot in the aquarium. Carry out this task on a regular basis.

Water changes

In week three we began a programme of water changes to dilute the constant production of nitrate. After the influx of algae in the aquarium, we were left with a residue of dead algae amongst the plant leaves and on the substrate. Again, it is vital to remove this dead matter to prevent it rotting down in the aquarium, and the best tool for this job is a siphon-action gravel cleaner. Here we look more closely at using this device and making a water change at the same time.

Most good models have a built-in, self-start priming mechanism so there is no need for you to suck on the end of a hose and risk taking in a mouthful of aquarium water. The broad tube and wide

foot allow you to 'hoover' a large area of the tank, as you siphon out the water into a bucket. Guide the foot gently over the plant leaves to remove the fine dusting of dead algae and also into the top of the substrate. This will be sucked up into the tube of the gravel cleaner and whirled round in the water flow to separate it from any debris. The heavier substrate particles then sink back down to the floor of the aquarium and the slightly more buoyant silt is sucked out and into the bucket. Repeat this process over the entire display, but make sure you do not remove more than a total of 25% of the aquarium water in the course of the cleaning operation. Do not worry if you do not manage to 'hoover' the whole tank; just remember to tackle the parts you missed the following week. As the bucket fills up with water from the aquarium, stop the siphoning and empty

Below: *Guide the siphon cleaner from one side of the aquarium to the other. It will remove organic debris in the substrate, as well as any waste matter disturbed while you trimmed the plants.*

Right: *Continue to siphon out the water and clean the plants and gravel until you have removed as much algae and debris as possible and the water is at the desired level. Do not siphon out any fish!*

A flexible hose attached to the siphon takes water from the aquarium.

Right: *Direct the water from the tank into a bucket on the ground. Siphon cleaners operate on the basis of removing lighter wastes while leaving the substrate relatively intact.*

it. Then restart the siphon as before. Do not discard the last bucket of water taken from the aquarium as you will need it for the next maintenance stage: the essential task of filter cleaning.

Cleaning the internal filter

Any filtration system, whether it be an external canister, hang-on or the internal filter used in this aquarium, needs regular maintenance if it is to function properly and have a long life. Filters can be separated out into three sections: the moving parts, the filter casing and the biological filter media. Start by dismantling the filter and clean each part as illustrated on the following pages. Place the pieces in the bucket and rinse them in the old aquarium water, using a small piece of filter wool to remove stubborn debris.

The only moving part in this unit is the impeller that spins to pump the water. Clean it and check it for wear and tear. If it shows any sign of damage after a few months, be sure to replace it, as a worn, off-centre impeller will undermine the motor head and render the unit useless. Impellers are a readily available spare part from your aquatic store.

The last part to be cleaned, because it is the dirtiest, is the biological filter medium. In the case of this tank, the filter medium is the blue foam, which also traps solid debris from the aquarium. Gently squeezing the foam in the bucket of old aquarium water will release nearly all the solid debris. The most important thing to remember at this stage is that the filter medium supports the vital *Nitrosomonas* and *Nitrobacter* bacteria on which the fishes' lives depend. NEVER wash any filter media in tapwater, which contains chlorine and chloramine that will kill the beneficial bacteria in the filter media (see page 92 for more details).

When all the parts of the filter are clean, reassemble the unit and return it to the aquarium. Remember to turn on the electricity, and make sure that all the systems are working properly. The filter will probably need cleaning every month.

Above: *Before carrying out any maintenance work on the tank, always switch off the electricity supply and, to be on the safe side, unplug it as well. Then start by removing the filter from its cradle.*

Right: *Working over the bucket of water that you siphoned out of the tank, remove the impeller and put it in the water. Stand the filter in the bucket while you work on each component.*

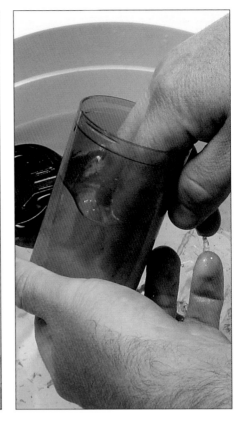

Above: Open up the casing that houses the filter media. Internal filters need cleaning every four to six weeks. If you wear rubber gloves, choose ones that allow you to use your hands freely.

Above: Remove the filter foam from the casing and put it in the bucket. Using water from the aquarium means you will not destroy the bacteria that have accumulated in the sponge.

Above: Wipe out the casing with filter wool dipped in aquarium water. When it is clean you can see through it more clearly so that you can monitor the state of the filter foam inside.

Above and Right: Using fresh filter wool if necessary, wipe the casing inside and out, taking care to remove any plant fragments that have become trapped in the narrow slots.

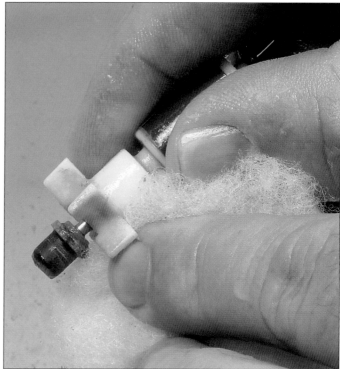

Above: *Clean the insert plate that holds the media in the filter. Because the aquarium relies so heavily on the efficiency of the filter, it is worth cleaning each element as thoroughly as possible.*

Above: *Clean off any slime that has accumulated on the impeller and the shaft, as well as in the housing. Take this opportunity to check whether the bearings need replacing.*

Activated carbon

Activated carbon is a very useful and widely available filter medium. It will remove chemicals, toxins and organic wastes by adsorbing them onto active sites on its surface. It can be prepared from a number of sources, and these give the carbon different characteristics. Using a blended product is the best strategy. It should be housed in a separate chamber in the filter or contained in a filter media bag. This is because it will have to be removed if disease treatments are used in the aquarium. Replace the activated carbon every four to six weeks, in line with the maker's instructions.

Below: *If the filter has a separate media compartment, you can use it for activated carbon. Before refilling it, clean the inside and rinse out the foam pad at the bottom.*

Above: *Before use, rinse new batches of activated carbon in a fine net held under running tapwater. This will flush out any carbon dust and small particles that would otherwise cloud the aquarium.*

Below and right: *Remove the foam from the canister and squeeze it out in the aquarium water in the bucket to retain the colony of beneficial bacteria. Remove any visible debris.*

Below: *By the time you have finished, the filter foam will look a lot cleaner. It is a good idea to replace the filter foam when it shows signs of not returning to its original shape when squeezed.*

When renewing filter foam, cut the old and new pieces in half. Run the new and old halves together for a month to allow the new filter foam to become seeded with the nitrifying bacteria.

Left: *Reassemble the filter in the reverse order of dismantling it. Ensure the foam sits properly in the canister as water will take any opportunity to bypass it, making the filter inefficient.*

Right: *With the impeller in place, fit the water pump securely on top of the canister. Replace the unit in the tank and ensure that it floods fully. Any air trapped inside will stop the filter working properly.*

Do not allow the filter medium to dry out during cleaning. Return the unit to the tank as soon as possible.

You can take advantage of the lower water level to carry out some more serious plant maintenance. In the righthand corner of this aquarium, the *Echinodorus* 'Red Special' and *Cryptocoryne wendtii* 'Tropica' have grown vigorously and are overshadowing the *Glossostigma* in the foreground, even though we left room for all the plants to develop. Without sufficient light, the *Glossostigma* will suffer. To prevent permanent damage to it, there are two options: either prune back the taller midground plant or move the foreground one. Start by examining the midground plant. Will pruning it give it an unnatural appearance? If so, then

moving the foreground plant is the best option. On this occasion, we have decided to move the foreground plant. First identify where it is to go. Do this before you put your hands in the tank so that you can see where the light penetrates to the base of the aquarium. Once you have identified the new planting position, cradle your fingers and push them very gently into the substrate all the way around the base of the plant. Slowly lift the plant free of the substrate. Replant it with your hand in the same cradle position and coax the substrate around the base of the plant. Once in its new position, it will start to flourish after a couple of weeks.

Refilling the tank

When all the maintenance is complete, it only remains to refill the tank. Ideally, mix together warm and cold tapwater in a bucket so that the temperature is roughly the same as that in the tank. Add a tapwater conditioner as described on page 92. When this has been thoroughly incorporated, pour the water into the aquarium using a jug. It is not a good idea to pour the water straight from the bucket, as it is heavy and difficult to control. A huge wave of water poured into the tank in one go can easily disrupt your display by uprooting plants and dislodging the carefully positioned decor.

1 *This glossostigma has become overshadowed by the plants around it and needs to be moved forward.*

2 *Holding the rootball, loosen the substrate around the plant and lift it out gently. Try not to damage the delicate roots.*

3 *Identify the new planting position and make a dip in the substrate with one finger. Bury the plant roots in the hole.*

4 *Bank up the substrate around the plant and firm it in. It is important to monitor the progress of all the plants in the aquarium, especially the smaller varieties.*

5 *In its new, brighter position, the glossostigma should make good progress. A living aquarium is a changing environment and adjustments like this help to keep it looking its best.*

Giving overshadowed plants a chance

Left: *Having moved one Glossostigma elatinoides, look at the others to see how they are progressing.*

This one at the righthand side of the aquarium is also overshadowed.

Left: *Once relocated, this plant is in a much better position. In the right conditions, the three plants will soon grow together.*

Left: *When all the maintenance has been completed, refill the tank with water that has been conditioned as described on page 92. Even small amounts of untreated tapwater are harmful to fish.*

Above: *Pour the conditioned water slowly into the aquarium to avoid disturbing the plants and substrate. Ideally, it should be at or near the same temperature as the water in the tank.*

157

FEEDING PLANTS

Like all living organisms, plants need food in order to grow and remain healthy. The nutrients they require are usually described as macro- or micronutrients. Macronutrients are required in greater quantities than micronutrients, but both are equally important to the plant's survival. Macronutrients are mainly responsible for building the structure of plants while micronutrients are used in biochemical processes within plant cells. The roles these nutrients play are listed in the panels on these pages.

Sources of nutrients

Tapwater is usually well supplied with several of the macro- and micronutrients

Essential micronutrients

Boron is used in many processes, including flower and root production.

Chlorine is required in minute quantities for biochemical processes, including photosynthesis.

Copper is used by enzymes involved in respiration.

Iron is a vital trace nutrient required for chlorophyll synthesis, enzyme production and respiration.

Manganese activates enzymes in chlorophyll production and photosynthesis.

Molybdenum is used in an enzyme to break down nitrate into useful ammonium for building proteins.

Nickel is used in very small amounts for an enzyme that converts urea to ammonia.

Zinc is used for chlorophyll production. Only required in tiny amounts.

Right: Pour a measured dose of fertiliser into the water. Do not be tempted to add more than directed. Over-fertilisation can be just as harmful as not feeding enough.

Below: Good-quality liquid fertilisers can provide a valuable source of iron and other essential nutrients to aquarium plants. Use them regularly, usually every one or two weeks.

required by plants. Indeed, sometimes they are present to excess, as in the case of phosphate. Generally speaking, if you carry out routine partial water changes, you can be confident that your aquarium plants will receive a regular supply of the nutrients they need. A proprietary liquid fertiliser will supply and/or supplement any shortfall.

Nitrogen is available to plants both from tapwater and in the form of nitrate, the end product of the ongoing nitrogen cycle (see page 94). Plants can also absorb ammonium from the water as a source of nitrogen.

Terrestrial plants obtain their supply of carbon from the carbon dioxide gas (CO_2) in the air, but submerged aquatic plants rely on carbon dioxide dissolved in the water. This is never available in sufficient quantities in an aquarium for plants to grow and remain healthy, and so levels must be boosted by the aquarist for vigorous displays. Without sufficient CO_2, the amount of photosynthetic activity required for the plant to thrive cannot be achieved. The carbon dioxide dosing system installed in the aquarium during the initial setup (page 28) will provide the required level of CO_2 for this display.

The laterite mixed into the substrate when the aquarium was assembled forms a major food store for the plants, while the bacterial culture ensures that the nutrients are released in a form that the plants can readily use. The reddish colour of the laterite indicates the presence of iron, which is vital for healthy plant growth. The laterite will nourish the plants during the first year. Thereafter, additional fertilisation is needed to meet their nutritional needs.

Plants can most easily absorb iron in the 'smaller' bivalent form (Fe^{2+}) and this will be the form that aquarium fertilisers supply. Iron is one of several metal-based micronutrients that can be 'lost' to plants in an aquarium by being oxidised into a 'larger' form (in this case, Fe^{3+}) that plants find difficult to take up. In order for iron, for example, to remain continually available to aquarium plants, it should be supplied in a chelated form. Chelates are organic compounds that bind to iron and prevent it from being converted to the 'larger' form that cannot be taken up.

Although liquid fertilisers generally provide many of required micronutrients, they do not supply the major macronutrients. (Adding extra doses of nitrate and phosphate will only encourage unwanted algae.) Most liquid fertilisers should be supplied every seven to fourteen days, as the nutrients are quickly used up by the aquarium plants.

Tablet fertilisers are an alternative to liquid products. Unlike liquid feeds, which 'blanket feed' the whole aquarium, you can use tablets to feed particular plants. They are also ideal for feeding plants once the laterite in the substrate is exhausted after the first year. Tablets are usually packed full of iron and other trace elements and release their nutrients slowly. Always follow the manufacturer's dosage instructions. They are easy to use; simply push them into the substrate near the roots of each plant you wish to feed.

The final source of plant nutrition comes, perhaps unexpectedly, from fish foods. They are usually rich in both potassium and phosphate, and waste passed by fish feeding on the foods will provide plants with a steady supply of these nutrients. There is no way of accurately dosing the nutrients supplied in this way, but in an aquarium well stocked with fish this can be a good way of feeding the aquarium plants.

Uneaten fish food can also feed plants, as bacteria break down the leftover fragments and release the nutrients they contain. There is a balance to achieve here in not overloading the aquarium with excessive amounts of uneaten food that will ultimately pollute the water with ammonia and threaten fish health.

Essential macronutrients

Calcium is used in cell wall structure.

Carbon is used as a major building block of plants – in fact, of all living organisms.

Magnesium is a vital component of chlorophyll and is needed to activate vital enzymes involved in biological processes.

Nitrogen, mainly in the form of ammonium, is used to build up proteins.

Oxygen is used in cell wall structure and for respiration to release energy from food. It is the major by-product of photosynthesis.

Phosphorus is essential for healthy root formation and flower production. It also plays an important role in energy transfer processes within the plant.

Potassium is vital for key biological functions in growth and reproduction.

Sulphur is required for producing proteins and the photosynthetic pigment chlorophyll.

Adding a tablet fertiliser

1

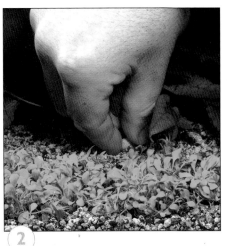

2

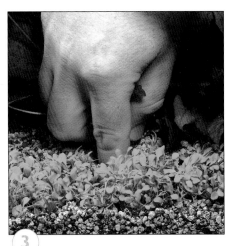

3

Tablet fertilisers supply nutrients directly to the roots. 'Greedy' plants that need plenty of iron will benefit from this method of fertilisation.

Tablet fertilisers are supplied in blister packs. Pop one tablet out of the pack and place it on the substrate near the roots of the plant.

Push the tablet into the substrate under or close to the roots so that they can easily absorb the essential nutrients as the tablet breaks down.

ADDING THIRD FISH

The final group of fish to be added to the aquarium are species that have more specific needs than the first two groups.

The pair of elegant cockatoo cichlids, *Apistogramma cacatuoides,* will form a lasting bond in the aquarium and may even breed. Their breeding behaviour involves setting up a territory to protect their young and to do this they will bully the other inhabitants. By introducing them last, they will not think the whole tank is their territory and will fit in more sociably with the other fish.

We are also adding some otocinclus catfish at this stage. They carry out a very important housekeeping role in the aquarium by eating algae. Ideally, there should be some algae for them to eat. Even though we try to create algae-free conditions, it is hard not to have an ongoing population somewhere in the aquarium. If algae are in very short supply, the otocinclus will soon learn to feed on the algae wafers fed to the other catfish.

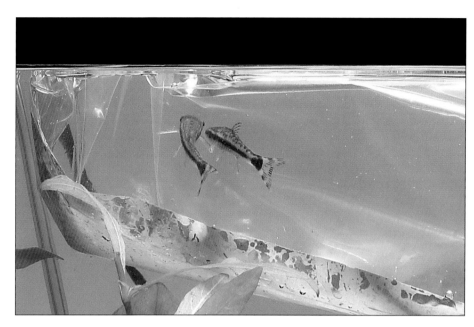

Above: *Now that the water conditions have stabilised, it is safe to add the third batch of fish. These are otocinclus catfish. Float the plastic bag in the aquarium for about 20 minutes.*

Below: *A pair of colourful cockatoo cichlids* (Apistogramma cacatuoides) *waiting to be released from their plastic bag, Ideally, quarantine new fish to safeguard your existing population.*

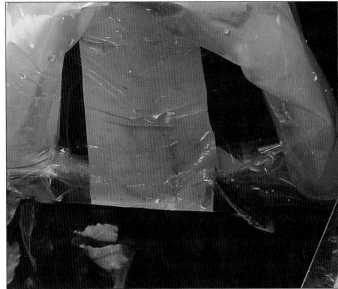

Above: When the water temperatures in the bag and in the aquarium have equalised, you can release the fish. Always check that there are no fish left in the bag before you dispose of it.

Above: Upend the bag just above the water surface and allow the fish to swim out. Do not tip them into the tank from a great height. This aquarium will house a group of otocinclus catfish.

Right: Unlike some of the cichlids, cockatoo cichlids (this is a male) can be safely housed in a planted aquarium. At 8cm (3.2in), they do not grow too large and will use the plants to establish territories. They may even breed.

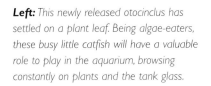

Left: This newly released otocinclus has settled on a plant leaf. Being algae-eaters, these busy little catfish will have a valuable role to play in the aquarium, browsing constantly on plants and the tank glass.

161

THIRD FISH

The third and final group of fish to add to the aquarium will generate the real 'wow' factor. They are best left till last, either because they require the security provided by established plants or because they are somewhat territorial. If added last, they will not regard the tank as their exclusive territory and try to chase off other fish added after them. These fish will probably be the most expensive ones you buy, so it pays to introduce them into a mature, stable environment that has been running successfully for several weeks.

Female

Male

The longer dorsal fin of the male black phantom tetra is clear, even in these young fish. The female (top fish) has reddish pelvic fins and does not develop such a large dorsal fin.

Black Phantom Tetra

Megalamphodus megalopterus

The beauty of this small tetra from Brazil lies in its silky black markings that put it in a class of its own. It is the complete opposite of the vivid neons and cardinals, and the 'silent' colours speak loudly when these shoaling fish are contrasted against their brighter cousins in the display. Although a firm favourite in many aquariums, it is not the easiest tetra to keep. Good water quality, including low nitrate levels, must be maintained if it is to thrive. In good conditions, the males may exhibit breeding behaviour and have mock fights. This should not be confused with bullying and is usually harmless to the individuals involved. Do not keep black phantom tetras with larger, potentially aggressive fish. Maximum adult size: 4.5cm (1.8in).

Rummy Nose Tetra

Hemigrammus bleheri

The rummy nose, from Colombia and Brazil, is one of Nature's little stunners and a real gem for any aquarium. Its black-and-white striped tail and silver body with a bright red marking running from the nose to behind the eye look as if they had been painted by a child. Its sleek, torpedolike body is suited to life in fast-flowing streams and the rummy nose will enjoy playing in the flow return from the filter. Pay particular attention to water quality; even high nitrate levels will cause this delicate tetra to suffer. The rummy nose does best in a well-planted aquarium where the water quality is at its best and there is ample cover. A minimum of eight individuals will make a good display. Avoid keeping them with larger, potentially aggressive fish. Maximum adult size: 4.5cm (1.8in).

Red Phantom Tetra

Megalamphodus sweglesi

The red phantom tetra from Colombia is a close cousin of the black phantom, featuring similar black markings but on a red background. It is even more delicate than the black phantom and requires very good conditions to do well. However, the effort is worthwhile, as the fish will reward you with its stunning red hues and placid behaviour. The red phantom should be one of the last fish you add to your display, as all the aquarium conditions need to be as stable as possible before you introduce it. The red phantom looks best in a small shoal of six to eight individuals. Avoid keeping it with larger, potentially aggressive fish. Maximum adult size: 4.5cm (1.8in)

CICHLIDS

The cichlids in your aquarium will form the real 'personalities' in the display. They are some of the most colourful fish you can choose, but restrict yourself to a pair from a single species for all but the largest of community displays. Males and females are usually very different in many ways – colour, body shape, size – and this adds to the attraction of this group of fishes.

In a healthy, well-maintained aquarium, many of the cichlid species on these pages will readily spawn and produce large quantities of fry. Some of the young will be predated on by the other fish, as would happen in the wild. Only a few healthy and lucky fish will grow to adulthood, but this all adds to the excitement of fishkeeping. Should your cichlids spawn they will defend an area of the tank as their own, but as long as there is plenty of foliage for the other fish to hide in, and the cichlids are not housed with overly slow species, then all the tank's inhabitants should have room to coexist. All of the cichlids described here should be kept with peaceful species only.

Laetacara dorsiger

These lovely round-faced cichlids from Bolivia and Rio Paraguay bring welcome colour into the aquarium. Their red bellies, with hints of blue when the males and females are breeding, show off the fish well. Both the male and female have a characteristic black spot on the dorsal fin, but the male's dorsal fin is much more developed. Like most small cichlids, they will be busy around the aquarium and have a huge appetite. They prefer some form of cover in the display, such as bogwood or rocks. This little cichlid is very peaceful, even when spawning, so makes an ideal fish for any community display. Do not confuse this species with *Laetacara curviceps,* whose aggressive nature during breeding makes it unsuitable for the community aquarium. Maximum adult size: 6-8cm (2.4-3.2in).

Keyhole Cichlid

Aequidens maronii

Because this small, peaceful Guyanan fish mixes well with most other species, it is one of the most commonly kept community aquarium cichlids. The distinctive 'keyhole' marking on each flank is vivid black when the fish is happy, but in stressed fish the colour of the flank fades to a murky brown all over. Keyholes also have a black line running through the eye down each flank to the edge of the gill cover. These cichlids thrive best when kept in pairs, and will lovingly care for their fry for a few months before allowing them to fend for themselves. The fish will benefit from relatively dense planting to provide cover in the aquarium, which they use for their own security. Maximum adult size: 10-12cm (4-4.7in).

Agassiz's Dwarf Cichlid

Apistogramma agassizii

This small, but stunningly elegant, Amazonian cichlid makes up in colour for what it lacks in size. The beautiful golden-red back and dorsal fin are in complete contrast to the petrol-blue flanks and tail. The blue coloration also spreads in a marbled pattern onto each side of the head. Males are generally larger than females and have a more pointed tail. These fish are often sold in pairs. However, if they are to breed, you will need to house one male with several females. Being so small, Agassiz's cichlid appreciates the cover provided by dense planting and root-shaped pieces of bogwood. They prefer slightly acidic, soft water. Maximum adult size: 8cm (3.2in).

Viejita Dwarf Cichlid

Apistogramma viejita

In this colourful little South American cichlid species, the male's bright yellow breeding colour is simply stunning. He is much larger than the female, with elongated fins and black spots along the back and flanks. These fish may spawn in your aquarium and will take over a cave in order to do so. They will defend this from the other inhabitants in the aquarium, but only if they approach too close to the eggs or fry. *A. viejita* will thrive on a good-quality flake food. Maximum adult size: Males 7.5cm (3in), females 3-4cm (1.2-1.6in).

Borelli's Dwarf Cichlid

Apistogramma borellii

The flowing fins of this elegant small cichlid often look a bit big for its body! They carry the petrol-blue and gold colours from the body and create a wonderful display. The males are half as big again as the females and have pointed dorsal and anal fins. These South American cichlids like to spend time amongst leaves in the upper layers of the aquarium, so include taller plant species in your display. Maximum adult size: Males 8cm (3.2in), females 4-5cm (1.6-2in).

Checkerboard Cichlid

Dicrossus filamentosus

The small, delicate stature of this superb South American cichlid makes it suitable for the smallest aquarium display. The checkerboard spots along each flank are bordered by two light blue lines that end on the pointed tail. The male's fins turn red as he enters breeding condition, while those of the female remain transparent. Both sexes enjoy the cover of a densely planted aquarium. If a pair should spawn they will defend the nest, but most other fish will be able to keep out of their way. The checkerboard cichlid will fit in with the majority of fishes, but avoid keeping it with other cichlids. It needs soft, acidic water. Maximum adult size: Males 9cm (3.5in), females up to 6cm (2.4in).

Cockatoo Cichlid

Apistogramma cacatuoides

Both the male and female cockatoo cichlid from the Amazon have stunning coloration, which makes them striking show fish for any community aquarium display. The male sports a huge, pointed dorsal fin, patterned with red, orange and yellow, hence the fish's common name. The smaller female contributes just as much to the display, bright yellow when spawning, with a jet black line along each flank. Despite her smaller size, the female can be the more adventurous fish as she moves around the aquarium. The cockatoo cichlid is difficult to breed, as males prefer to have several females in the aquarium, which can ruin the balance of your display. However, a single female may choose a breeding cave amongst rocks and wood in the aquarium and the male may succumb to her charms. Avoid keeping with other cichlids, unless they are in a large aquarium. Maximum adult size: Males up to 9cm (3.5in), females up to 5cm (2in).

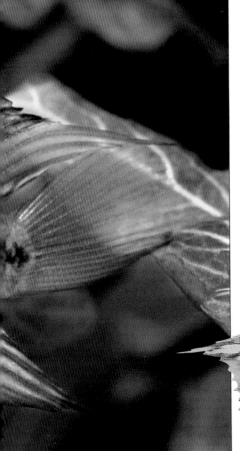

Kribensis

Pelvicachromis pulcher

The krib, from Nigeria, is the community cichlid most commonly found in aquarium shops. This popularity is easily explained by the fish's spectacular colours. The head is often striped black-and-yellow, while the underside has a vibrant red glow. As with all small cichlids, the colours become even more marked if the conditions are suitable for breeding and the fish start to court. Males can be identified by their pointed dorsal fin and they also tend to be larger than the females. This is one species that will readily breed in the aquarium. The female lays her eggs on the roof of a cave or on any flat surface. The fry hatch within a few days and will be guarded by the parents until they are old enough to forage for food by themselves. To give the fry a good start in life, feed them with a liquid fry food. This will supply all their needs until they can take solid fry food. Only keep one pair of adults per aquarium. Maximum adult size: 8-10cm (3.2-4in).

Female ram

Male ram

Striped Dwarf Cichlid

Pelvicachromis taeniatus

Although similar to the kribensis, *P. taeniatus*, from Nigeria and Cameroon, is usually more yellow in colour. However, this species can have many different colour variants, depending on where the original wild-caught specimens came from. The red body spot can range from yellow-red through pink to purple. A common feature on most colour variants is a gold tail with black spots. *P. taeniatus* will take advantage of plant cover in order to feel secure, but does benefit from a generous space in which to swim around freely. Although these cichlids can be bred in the aquarium, they are more difficult to raise than kribensis, as the fry are harder to feed. However, this should not deter you from keeping these vibrant little cichlids. Keep in slightly acid water. Maximum adult size: 7-9cm (2.75-3.5in).

Ram, or Butterfly Cichlid

Microgeophagus ramirezi

The ram's bright blue, reticulated scales create a stunning backdrop for the black, white and gold patterns towards the head of both the male and female. These stunning colours are set off by a red outline to both the top and bottom of the tail. The relatively small size and very peaceful nature of the Ram, which comes from Venezuela and Colombia, make it ideal for the smaller community display. Prefers some plant cover, and soft water. Maximum adult size: 7cm (2.75in).

Nanochromis parilus

Although it will venture up to midwater level, especially when food is on offer, this subtly coloured small cichlid is very much at home in the lower reaches, where it enjoys scooting around the bottom of aquarium. The elongated body adds a different shape to the display. It is yellow-brown in colour, with a blue-and-red underside. The tail and dorsal fin are brightly marked with stripes and dashes. These cichlids from Zaire are best suited to the larger display, as males (top fish) can be a little territorial. An aquarium with plenty of plants and decor will provide cover, both for the cichlid and its tankmates. It needs very clean, soft and slightly acidic water. Maximum adult size: 7-8cm (2.75-3.2in).

Small cichlids to avoid

The following common cichlid species should not be included in the average community display as they have a tendency to be too territorial and therefore aggressive towards their tankmates. This does not mean that you should always avoid these fish, as many make excellent aquarium subjects. They do, however, require a more specific setup tailored to their needs. In addition to considering this shortlist, you should always take advice from your retailer regarding the suitability of any fish.

Firemouth cichlid *Thorichthys meeki*
Crown jewel cichlid *Hemichromis cristatus*
Convict cichlid *Cichlasoma nigrofasciatum*
Blue acara *Aequidens pulcher*
Port acara *Aequidens portalegrensis*
Festivum cichlid *Mesonauta festiva*
Severum *Heros severus*

Nijsseni's Dwarf Cichlid

Apistogramma nijsseni

This attractive fish from Peru is one of the best dwarf cichlids you can buy for your aquarium. It is an ideal community fish as it is rarely territorial, although pairs can sometimes be antagonistic. The male and female have very different but equally vibrant colours. The male has cobalt blue flanks, with a gold belly and pelvic fins, plus a red-rimmed tail. The bright gold female is much smaller, with black markings. Both fish need a well-planted tank to feel secure. Breeding them is difficult, but if the conditions suit the fish, they could spawn in your display. Maximum adult size: males 4-6.5cm.

Angelfish

Pterophyllum scalare

Angelfish have been a firm favourite with fishkeepers for generations. Their extended dorsal and anal fins create a diamond shape that is unique in the aquarium. In recent years, hobbyists and commercial farmers worldwide have bred many colour morphs, but all with the same distinctive body shape. The fish originate from the Amazon River, but the natural colour – silver body with black vertical stripes – is rarely seen, even though it is one of the most stunning combinations. The deep-bodied altum angels *(Pterophyllum altum)* look fantastic when kept in a shoal and have more vivid markings than their more commonly seen cousins. Angelfish have a reputation for picking off small tankmates, especially neon and cardinal tetras. Therefore, you should only keep them with peaceful community fish (other than small tetras) that are more than 3cm (1.2in) long. Angelfish will breed, but only in soft, acidic water maintained to the highest standards. Maximum adult size: 15cm (6in). (Altum angels up to 38cm/15in tall and 25cm/10in long.)

Angelfish (Pterophyllum scalare)

Altum angelfish (Pterophyllum altum).

CATFISHES

Clown Plec
Peckoltia vittata

This is the ideal suckermouth catfish for all but the smallest of displays and beginners will find it easy to cater for. It will happily browse on any algae growing on plant leaves and decor. In addition to tablet and wafer foods, it will relish the treat of a chunk of cucumber weighted down in the tank. The yellow-gold striped pattern adds another dimension to the display. In common with other suckermouth catfishes, it appreciates somewhere to hide. Maximum adult size: 10cm (4in).

Bristlenose Ancistrus
Ancistrus temminckii

The bristlenose ancistrus is a superb example of a suckermouth catfish. The head and body have a mottled brown-and-yellow pattern all over. However, it is the bristles on the nose that are so distinctive. They help this bottom-dwelling fish find its way around in the dark or when water visibility is poor, and are a wonderful example of the diverse body shapes that can add interest to your display. Bristlenoses often hop around on the aquarium decor, rasping any algae growth from rocks and wood. Provide them with some hiding places. Maximum adult size: 12cm (4.7in).

Peckoltia **L134**

This stunning little catfish from South America is one of many species available to the aquarium trade that has not yet been scientifically named, hence the 'L' number. It will thrive in a community setup, but can be a little territorial towards other catfish around its favourite hiding place. Like some other small catfish, it is omnivorous and will prefer a mixture of bloodworm and algae wafers in its diet. Check with your retailer about the water conditions it has been kept in; the pH should be no higher than 7.4. This wonderful little catfish grows to 11cm (4.3in). Provide an aquascape containing rocks and wood for this suckermouth catfish to hop around on.

Starry Bristlenose Ancistrus

Ancistrus sp. L59 'White Seam'

Starry white spots on black skin, plus white-edged dorsal and caudal fins combine to produce a truly distinctive fish. The peaceful nature of this bottom-dwelling Ancistrus is another reason why this little gem of a catfish is ideal for the community tank. Make sure it has some hiding places to which it can retreat from the brightest lights. This ancistrus prefers water on the slightly acidic side, but your retailer may stock some fish acclimatised to your local water conditions. Maximum adult size: 8cm (3.2in).

Glass Catfish
Kryptopterus bicirrhis

The almost translucent body of the aptly named glass catfish makes this an interesting and unusual Southeast Asian species. The backbone and supporting bones are clearly visible as it swims peacefully in the midwater level of the aquarium. Long barbels are another defining feature. Provide some cover with dense planting. Maximum adult size: 9cm (3.5in).

Dwarf Otocinclus
Otocinclus affinis

If your aquarium ever experiences unwanted algae growth, then the *Otocinclus* species are Nature's solution to the problem. Perched on even the smallest plant leaves or up against the glass, they rasp on the algae, keeping it under control at all levels in the tank. Given this constant browsing behaviour, it is important to observe the fish carefully and provide a supply of algae tablets and wafers when natural levels get low. These tiny fish are suited to displays of all sizes. Maximum adult size: 4cm (1.6in).

The popular, algae-eating *Otocinclus* species, of which *O. vittatus* is another example, are much better suited to a display than the traditional algae loach (*Gyrinocheilus aymonieri*), which usually eats plants as well as algae, and grows aggressive in the aquarium. The placid *Otocinclus*, on the other hand, minds it own business and grazes the entire aquarium of any algae (except blue-green, brown and brush), on whatever it may be growing. Other *Otocinclus* species are available and all require the same care. This means offering additional supplies of processed algae foods when natural levels are reduced. Maximum adult size: 4.5cm (1.8in).

The striking pattern looks dramatic in the aquarium, but helps to camouflage this fish in the dappled shade of its native waters.

Zebra or Plec
Hypancistrus zebra

The stunning little zebra plec is a relatively recent introduction and a real star of the fish world. In recent years, it has become more affordable and has rightly found its way into many aquarium displays. It is the contrasting black-and-white stripes that set it apart from all other species of catfish. Algae forms only a small part of the zebra plec's diet, so do not rely on this species to clear the algae in your aquarium. Fish food tablets and frozen or freeze-dried bloodworm and tubifex will suit this bottom-dwelling catfish from Brazil, but make sure it gets its fair share of the food on offer. Maximum adult size: 7.5cm (3in).

Twig Catfish
Farlowella vittata

The whiptail catfish from Brazil is an unusual addition to the aquarium. This 'stick insect' of the fish world looks streamlined and elegant perched on a stone, leaf or root, but its swimming action is somewhat ungainly. This is partly due to its body shape, but its movement is also restricted by protective armour-plated scales and it never swims far. These peaceful fish occupy the bottom level and prefer a tank interior with some hiding places, where they can avoid the more boisterous members of the display. So that they can feed in relative peace, offer them algae tablets and wafers in an area of the tank away from where the other fishes are feeding on flake food. Maximum adult size: 15cm (6in).

Whiptail Catfish

Sturisoma aureum

The whiptail catfish holds its pectoral and dorsal fins erect for most of the time. It is easily recognised by the broad, dark stripe running along the body from the nose to the evenly forked tail. The tip of the nose is slightly upturned, while the mouth is on the underside, halfway between the nose and the eyes. The fish is a rampant algae browser, controlling most unwanted growths to the extent that it will need dietary supplements in the form of algae tablets and wafers. The whiptail, from Colombia, is a peaceful fish that occupies the bottom and middle levels of the tank. If you direct the flow from the power filter so that it passes directly over an exposed piece of wood or rock, the fish will happily sit there. *S. aureum* is ideal for any aquarium over 50cm (20in) long.
Maximum adult size: 15cm (6in).

LOACHES

Hill Stream Loach

Gastromyzon ctenocephalus

The natural habitat of the Hong Kong loach is in fast-flowing rivers and streams, where it uses its disc-shaped pectoral and pelvic fins to grasp the rocks or roots on which it perches to prevent itself from being swept away. In the aquarium, it will enjoy sitting in the filter outflow and appreciates some hiding places. Being an algae-feeder, this useful member of the carp family browses the decor and plant leaves looking for food. Many similar-looking species from fast-flowing waters are offered for sale, especially members of the *Beaufortia* genus.
Maximum adult size: 5cm (2in).

Horse-face Loach
Acanthopsis choirorhynchus

This unique member of the carp family is a joy to keep. The horse-face loach is a long, thin fish that rummages beneath the surface of the substrate looking for food. This habit makes the fish a perfect ally for the fishkeeper, as it keeps the substrate clean and prevents it stagnating and becoming compacted. Often, the Horse-face loach buries into the substrate, leaving only its head showing, while all the other fish swim around oblivious to it. However, it is important to provide a substrate with no sharp edges on which the fish could damage itself. The horse-face loach accepts tablet and wafer food and enjoys a treat of freeze-dried bloodworm and tubifex. Its placid nature makes it ideal for the community aquarium. Maximum adult size: 23cm (9in).

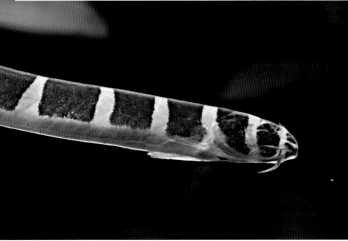

Kuhli loach
Pangio kuhlii

The kuhli loach from Southeast Asia is a long-standing aquarium favourite. Although the elongated body resembles that of an eel, it is a member of the carp family. The main reason for its appeal is its striking pattern of vivid orange-yellow spots or stripes on a matt black body. The kuhli loach is a notorious escapologist, so make sure your tank has a tight-fitting lid. Being nocturnal, it will hide during the brightest part of the day, emerging when aquarium light levels are reduced in the evening. Be sure to provide some hiding places so that this bottom-dwelling loach can take refuge during the day. It accepts crushed flake food and enjoys a treat of freeze-dried live foods, such as bloodworm and tubifex. Maximum adult size: 12cm (4.7in).

177

RAINBOWFISH

Very few aquarium fish come from Australia, but rainbowfishes are the exception. Many species are also found in the surrounding Southeast Asian region. As they become increasingly common in aquarium shops, their popularity is rising. This group of fish have a unique body shape and introduce a metallic colour accent not found in many other aquarium fish. Other attractions include their placid nature and active lifestyle – both ideal traits in community fish. They are best kept in a small shoal to help them feel secure, and make a bold centrepiece in the display. In fact, an aquarium can easily be turned over to these fish, as they capture interest at the expense of other species. They will feed on tablet and wafer food, with treats of freeze-dried bloodworms and brineshrimp.

Red Rainbowfish
Glossolepis incisus

Males of this midwater species have a strangely small head in relation to the huge body with a distinctive high back. Whereas males are a stunning red, females are more yellow and do not have the high back. Although it is an active fish it has a placid nature, which allows it to mix with all other community fish, making it the star of many displays. Provide some cover with dense planting and check your water chemistry before buying these fish; they prefer slightly alkaline water, with a reasonable level of hardness. If the water is soft, you should add some hardness minerals and salts. Do not keep it in small displays owing to its mature adult size. Maximum adult size: 15cm (6in).

Threadfin Rainbowfish
Iriatherina werneri

A unique display of fins give rise to this rainbowfish's common name. In the male, the double dorsal fin has two distinct shapes: the front fin is round and lobe-shaped, while the rear fin is tapered and threadlike. The rear dorsal fin is mirrored by an extended anal fin. Both are jet black and extend beyond the red-edged tail. These fin extensions are absent in the females. Being one of the smallest rainbowfish species, the threadfin is suited to any size of display. It occupies the middle and upper layers, where it needs access to dense plant cover. Do not keep it with Siamese fighters or boisterous barbs that will nip the males' extended fins. Maximum adult size: 5cm (2in).

When displaying, these young males will raise the long dorsal fin like a flag.

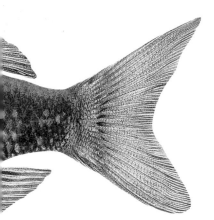

Boesman's Rainbowfish

Melanotaenia boesmani

The stunning colours of Boesman's Rainbowfish from New Guinea have made it a firm favourite with many aquarium-keepers. In adult fish, the head and front half of the body are blue, in stark contrast to the rear half, which is a mixture of red and gold. Do not be deterred from buying the fish you see in aquarium shops; they will probably be juveniles with only a hint of the coloration that will develop with maturity. This peaceful, midwater fish appreciates the cover provided by dense planting, but is less sensitive than other rainbowfishes to water conditions. Nevertheless, be sure to maintain a stable pH above 7. Maximum adult size: 10cm (4in).

Dwarf Neon Rainbowfish

Melanotaenia praecox

Metallic blue flanks and bright red dorsal, anal and caudal fins make this fish a real show-stopper. Like other rainbowfishes, it has a placid nature and enjoys the company of its own species. The flow from the power filter will provide a current in which this lively, midwater-swimming fish from Australia can play. Be sure to provide some plant cover. *M. praecox* prefers slightly acidic conditions, pH 6-7.5, and can tolerate reasonably soft water. Maximum adult size: 6cm (2.4in).

Western Rainbowfish

Melanotaenia splendida australis

This vividly coloured member of the rainbowfish family does best when housed in a larger display with four or five others of the same species. Like most rainbowfish it prefers plenty of open swimming space, but otherwise is one of the less demanding species. Feed them on a good-quality flake or granular food. Young specimens are a plain silver colour and it may take up to a year for the full colours to develop, but the wait is worth it. The red and blue flanks are a great contrast to other rainbowfish. Maximum adult size: males 10cm (4in); females 8cm (3.2in).

Banded Rainbowfish

Melanotaenia trifasciata

This multicoloured member of the rainbowfish family makes a superb addition to a medium-sized display. The overall colours of the fish are subject to regional variation and range from gold-green to silver-blue. However, regardless of body colour, all the fish have a black stripe that runs from head to tail, and red dorsal, anal and caudal fins. They are placid midwater swimmers that appreciate some dense plant cover. Maximum adult size: 12cm (4.7in).

CHARACINS

Three-lined Pencilfish

Nannostomus trifasciatus

Pencilfish are a timid group that bring a sense of calm to the aquarium. Three-lined pencilfish have a wonderful set of markings based around three predominant horizontal stripes. Between the top two lines the body is gold, and at the base of each fin there is a scarlet red patch. Feed these midwater swimmers with flake food and provide them with some hiding places. Do not keep pencilfish with busy tankmates that will intimidate them. This midwater swimmer from Brazil will spend most of its time holding station amid the security of aquarium plants, making it easy to take a long look at this charming little fish. Maximum adult size: 10cm (4in).

Headstander

Anostomus ternetzi

Its unusual behaviour makes the headstander an interesting subject for the aquarium. As its name implies, it often hangs head down in the water. Provide some hiding places and tall stem plants to help this placid midwater swimmer feel secure, and offer it a diet of flake food. Maximum adult size: 16cm (6.2in).

Marbled Hatchetfish

Carnegiella strigata strigata

The marbled hatchetfish is the most common member of a group of fishes that spend their time just below the water surface looking for food. The upturned jaw indicates that they are surface feeders, and although they accept flake foods, you should supplement their diet with frozen or freeze-dried bloodworm and daphnia. Provide some surface cover of floating plants. The body shape of these very streamlined fish from Peru resembles the keel of a yacht and gives them excellent stability in fast-moving water. They can often be seen holding position in the flow from the filter without apparently moving at all. Maximum adult size: 4cm (1.6in).

One-lined, or Hockey Stick, Pencilfish

Nannobrycon unifasciatus

This delicate species from Brazil is ideal for the smallest of quiet community displays, where it will swim in midwater at a head-up angle. It has an interesting pattern, namely a black stripe along each flank that extends into the lower lobe of the tail only. Like all pencilfish, it is timid and will only do well in a well-planted tank with plenty of hiding places. House pencilfishes with placid tankmates, and always keep them in a shoal of at least five fish as they feel most at home when schooling. Offer a diet of flake food. Maximum adult size: 6cm (2.4in).

KILLIFISH & ANABANTIDS

Lyretail Killifish

Aphyosemion australe

Killifish belong to one of the more specialist groups of aquarium fish that are occasionally seen in mainstream aquarium shops. Most species require acidic water, which means they are not suitable for keeping with many other community species. *A. australe*, from West Africa, may be offered for sale in your aquarium shop, having been acclimatised to water above pH 7. If you do buy one, make sure your water chemistry at home is suitable. The pH level should be no higher than 7.5 and hardness up to 18°dH. If you can provide the right water conditions in a densely planted tank, the midwater-swimming killifish will reward you with stunning displays of vibrant colours and patterns, mostly displayed on a gold-toned body. Offer them a diet of flake food, plus freeze-dried bloodworms and brineshrimp. Maximum adult size: 6cm (2.4in).

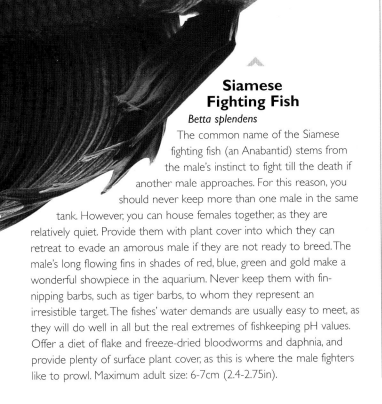

Siamese Fighting Fish

Betta splendens

The common name of the Siamese fighting fish (an Anabantid) stems from the male's instinct to fight till the death if another male approaches. For this reason, you should never keep more than one male in the same tank. However, you can house females together, as they are relatively quiet. Provide them with plant cover into which they can retreat to evade an amorous male if they are not ready to breed. The male's long flowing fins in shades of red, blue, green and gold make a wonderful showpiece in the aquarium. Never keep them with fin-nipping barbs, such as tiger barbs, to whom they represent an irresistible target. The fishes' water demands are usually easy to meet, as they will do well in all but the real extremes of fishkeeping pH values. Offer a diet of flake and freeze-dried bloodworms and daphnia, and provide plenty of surface plant cover, as this is where the male fighters like to prowl. Maximum adult size: 6-7cm (2.4-2.75in).

Fish to avoid

The following species are commonly offered for sale in aquarium shops but are not suitable as community fish.

Plecostomus, or plec *(Hypostomus punctatus)* This fish simply grows too large and destructive for a community display. A juvenile offered for sale is a tempting prospect but once it becomes a 30cm (12in)-long monster it can destroy your display. Far more suitable suckermouth catfish are available (see page 172-173). Another species to avoid for the same reason is gibbiceps plec *(Pterygoplichthys gibbiceps)*.

The orange-finned loach *(Botia modesta)* is often offered for sale as an alternative to the clown loach *(Botia macracantha)*. However, it is much more aggressive towards other fish and will harass more sedentary species until they become ill. The same can also be said for *Botia lohachata* and *Botia helodes*.

Two simple but important tips are never to buy any predatory fish, however tempting they may look, and also to take the professional advice of your local retailer regarding compatibility.

FEEDING FISH

Aquarium fish need a regular balanced diet if they are to thrive, and they depend totally on the fishkeeper to provide it.

In their natural habitat, many fish feed on live foods that form part of the natural food chain. Replicating this food chain is not a practical option, but good-quality, manufactured food will supply all the fishes' requirements. You can supplement this diet with 'safe' live foods in freeze-dried or frozen form.

Feeding your fish is one of the most useful tools for monitoring their health. Use this time to check that they are all coming out to feed and are in good physical condition. Careful observation will allow you to ensure that every fish gets its fair share of food.

Flakes and pellets

Generally speaking, the most practical food source for community fish is a proprietary flake or pellet food. Flakes are ideal for small to medium-sized fish, measuring up to 6cm (2.4in). Larger fish will do better on the more bulky, granular pellet food. The key to choosing any manufactured food is to look for a reliable brand that provides a compete food, including the following constituents.

Dried foods

Dried foods can provide the staple diet for most fish. Flake foods are the most common form and widely available.

Choose the appropriate foods for your fish. These sinking granules will suit bottom-feeding species.

Protein Fish need protein for growth and cell repair. It is also broken down to provide energy when required. Ideally, the protein should be provided by fish and/or shrimp meal, as these sources are easily assimilated without causing the fish to produce excess waste. For tropical aquarium fish, protein should constitute about 47% of the dry weight of the food.

Fat, or fatty acids Young fish use up more energy than adults and require a slightly higher proportion of fat in their diet to meet their needs. They use protein exclusively for growth. For adult fish, the fat content of the diet should be no greater than 6%. Any excess will go to waste and result in an oil slick on the surface of the water.

Vitamins All animals, including fish, need vitamins, and a deficiency can cause a variety of problems. Vitamin A is essential for growth, and Vitamin C for healthy skin. Poor skin condition is harmful for fish, as the mucus on the skin is a first line of defence against disease. Vitamin B_3 enables fish to process the protein in their food. Vitamin D_3 is essential for healthy bones; without it, fish may develop weak, deformed skeletons.

To help you make an informed choice, all the vitamins contained in the food should be clearly marked on the packaging. The levels indicated are those remaining in the food when it reaches its 'use by' date. Vitamins deteriorate quite quickly, so do not be tempted to offer old stocks of dried foods.

Left: *Flake foods are easy to use, but do not be tempted to overfeed. Offer a small amount at first; you can always add more later. In time, some fish may even take individual flakes from your fingers.*

Below: *Fish will soon respond to the arrival of food. This is a good opportunity to check that they are all present and feeding normally. Make sure that fish at all levels in the tank are getting food.*

Tablets and wafers

Tablet foods are much like flake foods, but in a different form.

Algae wafers are ideal for herbivorous algae-eaters and bottom-feeding catfish.

Left: *Tablets that are stuck onto the glass can be used to bring fish to the front of the tank for observation. As it disintegrates, the food attracts fish from all parts of the aquarium.*

Carbohydrates Generally, fish are better able to use fats and proteins, rather than carbohydrates, for their energy needs. A food may be high in energy, but if most of this is supplied as carbohydrate, the fish may not be able to use it as well. Carbohydrates are vegetable in origin and include complex sugars such as starch, as well as cellulose (fibre). Artificial diets have a reduced fibre content compared to natural diets.

Colour enhancers Certain ingredients harmlessly promote the natural skin colours of your aquarium fish, so that you see them at their best. Beta carotene and astaxanthin have long been used as effective colour enhancers, but have been surpassed by the introduction of Spirulina. This freshwater algae has very high colour enhancing properties and should be a standard ingredient in good-quality food.

Immune system stimulants This is a relatively new addition to aquarium fish foods. The immune system stimulant activates a fish's immune system so that it is ready to respond to any attack and does not need to wait for the attack itself to stimulate an immune response. This ensures that a fish is much more likely to fight off an infection successfully. The additive commonly used is Beta Glucan. An immune system stimulant should be a standard ingredient in modern aquarium fish food.

The right proportions of these ingredients are found in both flake and granular pellet foods. When buying new fish, it is always worth asking the retailer what type of food they have been accustomed to so that you can continue the same feeding regime in your home aquarium. Make any changes to the diet gradually.

Tablet foods

Tablet foods are usually made from standard flake food pressed into a tablet form. They have been developed to make it easier to watch the fish while they feed. Push the tablet firmly onto the inside of the front glass, where the water and the pecking action of the fish will cause it to break up slowly. The main benefit from this feeding method is that all the food is in one place and the fish must come to it to feed, which most species will readily do. Only offer tablets as an alternative to other foods; the fish will ignore one of the foods and much of it will rot in the water.

Wafers

A few species of fish, especially catfish and loaches, appreciate a high percentage of algae in their diet. Sinking wafers make it easier to supply the requirements of these bottom-dwelling fish. The hard wafers sink to the base of the aquarium, where they soften, so that the loaches and catfish can browse on them. It is a good idea to offer several smaller wafers, as some catfish can be quarrelsome over food and a timid individual may suffer if not given ample opportunity to feed.

Freeze-dried foods

Live foods, such as bloodworm, tubifex worms and daphnia, are sometimes offered for sale, but are best avoided as they often carry disease organisms that can easily infect your fish. Freeze-dried and frozen foods are a safe alternative. They are gamma-irradiated to kill fish pathogens and then freeze-dried to maintain the nutrient content of the food. Ideally, you should feed these foods to

your fish as a treat, in conjunction with a flake food diet. This gives the fish a variation in food texture and the most complete diet. Bloodworm, tubifex worms and daphnia are all popular freeze-dried foods. They are often premoulded into blocks to make the food easier to use.

Frozen food

Frozen foods are also gamma-irradiated to remove pathogens, and then packed into blister packs prior to freezing. Keep them frozen until they are required, but allow them to thaw before feeding them to your fish. Placing the required amount in a little warm water in a shallow dish will speed up the process. The only drawback of this food is keeping it frozen; this usually means storing it with food for

human consumption, which some family members might object to. Once the food has thawed never refreeze it. Either use it immediately or throw it away.

Live foods

As we have seen, water-based live foods can introduce disease into the aquarium, but this is not the case with terrestrial foods such as earthworms, which make a suitable treat for your fish if offered occasionally. Rinse them thoroughly to remove any soil and drop them into the aquarium. Be sure to choose worms of the right size in relation to the size of the fishes' mouths. Do not be tempted to place a large earthworm into the aquarium and expect the fish to nibble at it; they like to eat it whole. The worm will

Freeze-dried food

Freeze-dried mosquito larvae can form part of a more varied diet.

Freeze-dried tubifex is a safe way of offering this invertebrate food.

Frozen food

A defrosted cube of bloodworm makes a treat.

Holiday feeding

Aquarists are often concerned about how to feed their fish when they go away. There are various options. Healthy adult fish will happily go without food for a week or so. If you are away for longer than that, package up daily rations of flake or freeze-dried food in a twist of foil and leave them in the fridge for a friend or neighbour to administer. (Hide the tub of flake in case your helper is tempted to give the fish a bit extra.) Alternatively, invest in an autofeeder – a timer-controlled reservoir for flake or small granules – and programme it to dispense one or more meals daily. Vacation feeding blocks are another option.

Left: *Vacation feeding blocks contain food in a type of chalky block that gradually dissolves, allowing the fish to eat the feed it contains. Small sticks release the food over a weekend. Larger blocks last for a week or more.*

Right: *This battery-operated autofeeder can dispense two meals a day. At preset times, the food compartment rotates and the contents drop into the tank. Adjust the blue wheel to control the feed quantity.*

Right: *Daphnia, or water fleas, are microscopic creatures that make a good first food for newly hatched fry, as well as small fish. They are sometimes available in aquatic outlets.*

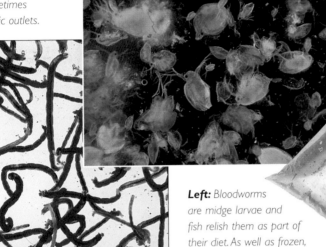

Below: *Suitable aquatic live foods include daphnia (left), bloodworms (centre) and brineshrimp (right). They are sold in plastic bags full of water. Strain the contents through a fine net. Do not put the water into the aquarium.*

Left: *Bloodworms are midge larvae and fish relish them as part of their diet. As well as frozen, they are available in freeze-dried form from aquatic stores.*

Right: *Adding a batch of bloodworm to the aquarium prompts an immediate feeding response from your aquarium fish. You can watch the fish feed at close quarters as they devour the larvae.*

simply crawl into the substrate, where it will die and rot, causing water quality problems. Remove any worms that remain uneaten after five minutes.

How much should I feed?

Overfeeding is the most common cause of water quality problems, which in turn lead to fish health issues. It is therefore vital to gauge how much food you feed your fish. The golden rule is to offer just a little at a time. With flake foods, start by offering a small pinch, say half-a-dozen

flakes. If the fish eat it all within five minutes, add a little more. If any is left floating or sinks to the bottom and is not investigated by the fish, remove it with your net straightaway. Uneaten food will decompose in the aquarium, causing a release of toxic ammonia into the water as it rots.

When should I feed the fish?

As feeding the fish is one of the few occasions when you can interact with them, choose a time when you can sit

down and enjoy watching them. Fish can become accustomed to a set feeding time, so choose a time that suits you. Feed them once a day for the first two months after setting up the aquarium. This will allow the biological system to establish without placing an excessive strain on the filtration bacteria. Thereafter, you can feed the fish twice daily. Always make sure that no-one else in the household has fed them first. This will avoid the risk of uneaten food being left to rot in the aquarium.

THE FINISHED DISPLAY

After twelve weeks, the aquarium is continuing to progress. Most things have turned out well, but the most difficult plants to grow – the *Hemianthus* and the *Glossostigma* in the foreground – are encountering some teething problems. They have not flourished as well as they should have. This could be the result of a number of factors: not enough light, not enough food, etc. The light issue can be resolved by adding a reflector to the front lamp in the aquarium. This will bounce all the light given out by the lamp down into the aquarium and increase the light intensity at the front, which should encourage the plants to put on more growth. If in time the plants show no signs of recovery, replace them before they rot in the aquarium, as we have done here.

The *Echinodorus cordifolius* is also showing less evidence of growth than before. This is because new plants will have been grown as an emergent plant, with leaves above water. Once the plant is completely submerged, it must produce submerged leaves as the emergent ones die away. Remove the emergent leaves, allowing room for the new submerged leaves to fill the space. This may take a few weeks. The continued CO_2 dosing and liquid feeding will create the correct conditions for the plants to thrive.

The biological filter has matured, with no signs of any ammonia and nitrite. Nitrate is being kept under control by regular water changes and the fish are proving the real stars of the show. Tablet food stuck to the inside of the glass encourages them to congregate in the central part of the aquarium, creating a dynamic kaleidoscope of colour in the heart of the display.

This is how the display aquarium looks with replacement plants for the Hemianthus *and* Glossostigma *in the foreground area.*

THIS IS JUST THE BEGINNING

The completed aquarium will now be the main attraction in the room chosen for it. In the twelve weeks of development, from selecting the rocks, wood and plants through to adding the fish, a bare glass tank has been transformed into a unique display to be proud of. With the wide range of decor, plants and fish available to hobbyists today, it is possible to create an almost endless variety of 'designs' in a home aquarium. One avenue to explore is to recreate natural environments from around the world, such as an Amazon pool, an African stream, a Southeast Asian swamp or a brackish estuary.

With proper care, the display will provide pleasure every day for years to come. It will be constantly changing, however, as plants grow larger or lose leaves before growing new ones. Some plants may flower or produce daughter plantlets on runners, and these will be ideal raw material for propagating new plants, as explained on page 199.

Below: *The fish will explore most areas of the aquarium, some enjoying the flow from the pumps, while others will avoid it. As the plants grow, they will sway in the water flow from the filter, adding more movement to the display.*

Below: *The male and female cockatoo cichlids have taken centre stage and will play out their courtship roles amongst the rocks and bogwood. Their breeding rituals add a new dimension to the display.*

Right: *The three 'windows' present different viewing angles into the display, allowing us to follow the fish as they swim between the plants. An aquarium is literally a living picture that will quickly become everyone's focus of attention.*

FISH HEALTH

To ensure that the fish stay healthy in our display aquarium, it is vital to carry out routine maintenance as described on pages 194-199. Should the water quality start to decline through lack of maintenance then we run the risk of fish succumbing to diseases that their immune system normally keeps at bay. Should this happen, we must be able to diagnose what our fish are suffering from and provide the appropriate treatment. The 'unfortunate' fish featured here has symptoms of all the common ailments that tropical freshwater fish may suffer from in the aquarium. Most are easily identifiable and can be cured using proprietary treatments. The key approach to monitoring the health of your fish is to be observant. Watching your fish as you feed them every day is the ideal opportunity to check their 'body language'. An unwell fish will not feed as well as usual or not at all and may swim around with clamped fins held tight to the body. Look closely at any fish behaving unnaturally for symptoms that match those shown here. This will help you identify the problem and apply the best remedy.

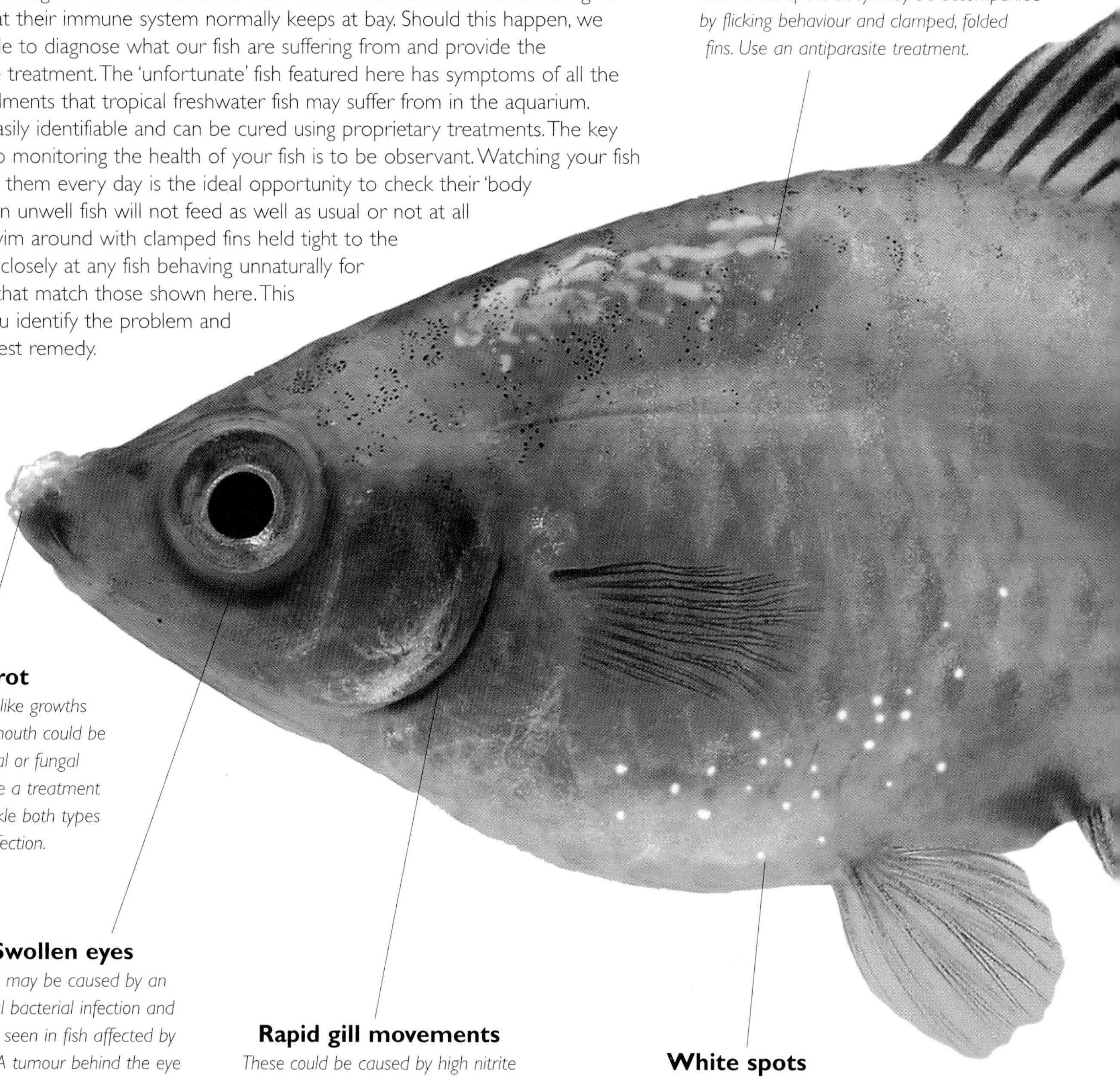

Slime patches

These are caused by the fish's immune response to protozoan skin parasites. In severe cases, the slime (mucus) patches will cover most of the body. May be accompanied by flicking behaviour and clamped, folded fins. Use an antiparasite treatment.

Mouth rot

Cottonwool-like growths around the mouth could be a bacterial or fungal infection. Use a treatment that will tackle both types of infection.

Swollen eyes

These may be caused by an internal bacterial infection and can be seen in fish affected by dropsy. A tumour behind the eye could also cause this symptom.

Rapid gill movements

These could be caused by high nitrite levels or other water quality problems, as well as parasites or bacterial infection. Check the water quality and make a change if necessary. Otherwise treat with a suitable medication.

White spots

White spots like sugar grains on the skin and gills are signs of a parasite infection called white spot, one of the most commonly seen problems among aquarium fish. Treatments target the free-swimming stages of the parasite's life cycle and to be effective must be carried out over a period of a few days. Left untreated, white spot will quickly spread to all the fish in the aquarium.

Fine gold spots

Very fine gold spots are typical symptoms of velvet disease caused by a skin parasite. Treat quickly with a targeted antiparasite remedy.

Fungus

Any skin damage may become infected with fungus to produce cottonwool-like growths. To be sure, use a medication that will deal with both external bacteria and fungus.

Ragged fins

Ragged edges to the fins, which may also be red and sore, are signs of a bacterial infection known as finrot. Use an antibacterial treatment to stop it spreading to the body. Damaged fin tissue may regrow once the infection has been cured.

Using medications

1 Carefully measure out the correct amount of treatment or the number of drops recommended and add this to a jug of water taken from the aquarium.

2 Mix the treatment into the water in the jug. Make sure that the spoon or other implement you use to mix the treatment is clean and free from any chemicals or other contamination.

3 Slowly pour the treatment over the surface of the aquarium. Diluting the mixture in this way will ensure an even distribution of the medication.

Important notes

Remove chemical filters before adding medication. Check that treatment is easy to use and has clear instructions on how to use any pipettes or measuring cups included.

Keep a note of the volume of your aquarium (see page 38) as this will allow you to dose accurately. Overdosing will probably be lethal to all your fish!

Always keep fish medication locked up out of the reach of children, as many contain potentially harmful chemicals.

Protruding scales

Swelling with scales protruding like a pinecone, commonly known as dropsy, is caused by an internal bacterial infection that inhibits the fish's ability to control the level of water in its body. Use an internal bacterial treatment.

Right: *A Cuming's barb with classic signs of dropsy. The whole body is swollen, making the head appear relatively small. The eyes are protruding. Difficult to treat, but if diagnosed early can be cured with proprietary treatments.*

ROUTINE MAINTENANCE

A LIVING DISPLAY

The display aquarium you have created is a living environment that will change and develop over a period of time. To keep it looking good, it is vital to carry out routine maintenance. All the living organisms in the aquarium produce waste. If this is allowed to build up, the biological cycles will become overloaded and the display will disintegrate. Routine maintenance tasks are presented here in easy-to-follow time slots. By following this regime you will be able to keep your display looking as stunning as it did when you added your last batch of fish.

Essential supplies

As well as making sure you have good supplies of fish food and plant fertilisers, it is vital to have the following supplies and spare parts:

Heater thermostat
Suckers
Bearings, impeller and 'O' ring
    seals for filters
Filter wool and or foam
Other filter media being used
    in the system
Clean plastic bucket/bowl
Activated carbon
Thermometer
Nets
Fuses
Fluorescent tubes
Starters for fluorescent tubes
Airline and airstones
Test kits
Tapwater conditioner
Filter start-up product

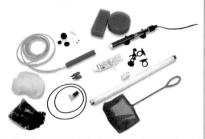

EVERY DAY

Checking your aquarium every day should be a pleasure, not a chore. As a general rule, the more often a task is needed, the less time it takes. Most of the daily tasks take only a few minutes.

Check the water temperature.

Check for missing fish. Dead fish left undiscovered in the aquarium will pollute the water and threaten the wellbeing of the other inhabitants. Also check for signs of ill-health or distress, such as red marks on the body and gills, excess mucus, gasping at the surface or any other unusual behaviour.

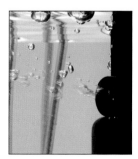

Check that the internal filter is working properly.

Feed the fish, making sure that they all receive some.

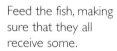

Check that the lights are working.

Remove any uneaten food that the fish do not seem interested in.

EVERY 7-14 DAYS

Among the routine maintenance tasks that you need to carry out every 7-14 days, the most important ones are geared towards keeping the water in good condition, such as regular water changes and tests for toxic waste products.

Test the water for pH, ammonia, nitrite and nitrate levels.

Gently disturb any fine-leaved plants to remove trapped detritus.

Clean the substrate with a gravel cleaner that will siphon off debris.

Clean the front and side glass of the aquarium to prevent a buildup of algae that can be very hard to remove later on.

Make a 15% water change and refill the aquarium with conditioned water at the same temperature.

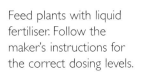

Feed plants with liquid fertiliser. Follow the maker's instructions for the correct dosing levels.

Where fitted, clean the condensation cover to avoid a reduction in light to plants.

Remove dead or dying leaves.

ROUTINE MAINTENANCE

EVERY 4-6 WEEKS

The maintenance tasks necessary every 4-6 weeks centre on the filter. This is the vital life-support system for the aquarium and needs regular attention to ensure that it is functioning properly. Any cleaning must be carried out with tank water to avoid disrupting the beneficial bacteria in the biological medium.

Clean the internal filter, including the impeller and casing.

If included, replace the activated carbon in the filter. Rinse it before use.

Clean the filter foam in tank water.

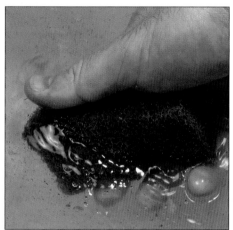

Renew the yeast-sugar solution in the CO_2 fertilisation system and clean the inlet of the water pump.

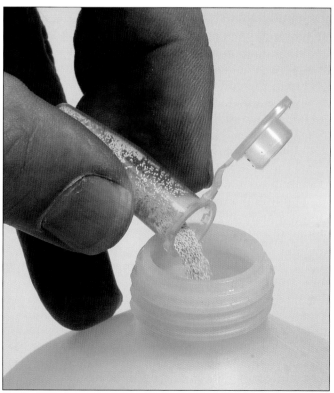

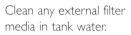
Clean any external filter media in tank water.

Replace expendable filter media, such as filter wool.

EVERY 6-12 MONTHS

The routine maintenance at this level focuses on replacing consumable pieces of aquarium equipment. Fluorescent tubes and parts of the filtration system do wear out, so to maintain optimum performance check and replace them before they fail completely.

Replace half the filter foam in the internal filter. Allow one month before replacing the other half

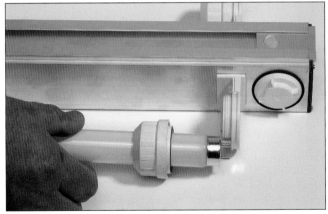

Replace fluorescent tubes, even if they are still working

Replace the filter pump impeller

AS NEEDED

If the aquarium has a full display of plants, keeping these healthy and growing well is one of the long-term objectives of routine maintenance. And if they outgrow their space, they will need trimming back. Always be prepared for emergencies and make sure you have spares of essential items in stock.

Replenish any tablet fertilisers for aquarium plants.

Trim tall plants to prevent them blocking light to other plants. Use trimmings for propagation.

Check the quarantine/hospital tank and make sure the filter is working properly in case the tank is needed for new fish or to treat existing stock with health remedies.

UNDERSTANDING ALGAE

Algae are simple plants that grow in the aquatic environment. They are usually out competed for food and light by the more 'advanced' plants in the aquarium. However, if they do appear you should take quick action to remove them before they ruin your display. Algae can grow very quickly, so swift action is vital. Always ensure that any algae treatment you use is plant friendly, as some may kill your plants as well as the algae.

There are various types of algae that may develop in the aquarium:

Green water algae will turn your aquarium into 'pea soup' as the algae bloom in the presence of too much light and nutrients. You can cut down excess light by altering your lighting timer or ensuring that no direct sunlight falls across the tank. Use a flocculent algae remover to remedy the problem.

Hair algae will grow like string from any internal surface in the aquarium, including the plants. The strands are usually green but can also be black or brown. Try and remove as much as possible by hand before adding a suitable treatment.

Experiment by shortening the aquarium lighting time to achieve a period that does not encourage the algae growth.

Slime algae appear to be similar to fine hair algae but they are species of bacteria. This green/brown carpet of slime can engulf an aquarium in a matter of a few days. It takes a hold only when water quality is poor and essential maintenance to reduce organic waste, such as removing uneaten food and dead leaves and gravel cleaning, has been missed. The best remedy is to carry out a water change, add a 'sludge control' antibacterial product to the aquarium and return to the regular maintenance regime. It will often die back as quickly as it appeared.

Right: *A glistening shroud of blue-green slime algae spreads rapidly over the fronds of an aquatic fern.*

Left: *Green water is a suspension of single algae cells that multiply rapidly in nutrient-rich, illuminated water.*

Below: *Neglected maintenance has allowed brown strands of hair algae to suffocate the plants they cover.*

The colour of water

Crystal-clear water may not necessarily be 'healthy', since it may contain toxic levels of ammonia and/or nitrite that represent invisible dangers to the livestock in your aquarium. However, there are some more tangible signs in the colour and condition of the tank water that can point to the following possible causes:

Yellow-stained water is 'old'. Missing water changes causes a build-up of organic dyes. The remedy is to make an immediate 20% water change, followed by a return to regular 15% water changes. Also add or change the activated carbon in the filter.

Brown 'tea-stained' water is usually caused by tannic acids leaching from bogwood that has not been soaked for long enough. If possible, remove the wood and soak it as recommended on page 30. Alternatively, increase the frequency of partial water changes. Some chemical filtration products will remove these dyes.

Brown / green / grey cloudy water is usually caused by overfeeding and lack of aquarium maintenance. Carry out a water change and clean the filter. Use a proprietary flocculent treatment to stick the particles together so they can be removed by the filter.

Above: *Tannins released from the bogwood have turned this water 'tea brown'. Not usually a threat to fish health, but unsightly.*

Propagating from cuttings

Trimming plants to tidy up the tank can provide raw material for propagation. Most stem plants can be propagated by cuttings taken from both the top and middle stem areas. Taking cuttings is also a way of 'thickening' up plants by stimulating side shoots.

1 To take a top cutting, snip off a length of stem with several leaves or nodes. Cut between the nodes with sharp scissors. For the best results, take cuttings from the fastest growing and/or healthiest stem.

2 Strip away the leaves from one or two nodes at the base of the cutting to allow the plant to root more quickly. Roots will form from the stripped node. Make sure it is beneath the substrate when planting.

3 Push the cutting into the substrate so that the lower leaves are just resting on the substrate surface. Roots should grow from the base and the plant will establish quickly, although lower leaves may die off.

Propagating from a runner

Many plants produce several daughter plants from one runner, so to obtain a larger number of plants, do not remove the runner until it has produced at least five or six plants.

The mother plant may produce more than one runner at a time.

*These daughter plants of **Echinodorus** sp. are ready to be planted in the substrate.*

2 Separate the individual plants, or 'slips', leaving a small length of runner on either side. Handle them carefully, holding them by the leaf and not by the stem.

1 Once the 'mother' plant has produced a number of daughter plants with at least two or three leaves each, you can cut the runner with a pair of sharp scissors.

3 Put each new plant into the substrate, leaving a gap of at least 5cm (2in) between the plants to allow for future growth.

GENERAL INDEX

Fish Index

FISH INDEX

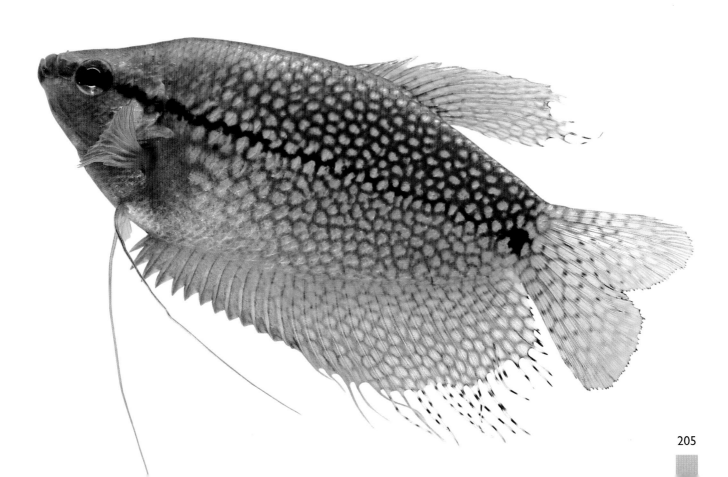

Plant index

A

B

C

D

E

F

G

H

L

CREDITS

Credits

The publishers would like to thank the following photographers for providing images, credited here by page number and position: (B) Bottom, (T) Top, (C) Centre, (BL) Bottom left, etc.

Aqua Press (M-P & C Piednoir): 11(B), 12(T), 60(L), 62(T), 64-65(BC), 65(R), 66(TR), 73(BR), 75(TC), 76-77(BC), 79(R), 80-81(C), 113(TL), 122(T,B), 123(B), 124(T,B), 124-125(B), 125(B), 126-127(B), 128(B), 130(C), 131(B), 132(B), 133(T), 134(B), 134-135(T), 135(T), 137(T,B), 138(CL,BR), 139(C), 162(C), 162-163(B), 164(B), 164-165(T), 166-167(B), 167(T), 168(T,B), 170-171(B), 172-173(C), 173, 176(L,T,C,B) 179(B), 180(B), 181(T), 182(B)

Jan-Eric Larsson-Rubenowitz 141(T)

Photomax (Max Gibbs): 172(T)

William A Tomey: 170(T,C)

Tropica (Ole Pedersen): 86(T)

Acknowledgments

The publishers would like to thank: Simon Williams at J & K Aquatics, Taunton, Somerset for supplying the fish featured in the display aquarium; Martin Pedersen of Tropica Aquarium Plants, A/S, Hjortshøj, Denmark, for supplying plants for photography; Kevin Chambers of Arcadia for supplying lighting equipment for photography; Keith Cocker of the Norwich Aquarist Society for help with fish photography; David Cummings of Kesgrave Tropicals for supplying aquarium equipment; Swallow Aquatics, Rayleigh, Essex for providing fish and onsite photographic facilities.

Index compiled by Amanda O'Neill.